Clinical Genetics:
Problems in Diagnosis
and Counseling

BIRTH DEFECTS INSTITUTE SYMPOSIA

Clinical Genetics: Problems in Diagnosis and Counseling

Edited by

ANN M. WILLEY
THOMAS P. CARTER
SALLY KELLY
IAN H. PORTER

Birth Defects Institute
Center for Laboratories and Research
New York State Department of Health
Albany, New York

ACADEMIC PRESS   1982
A Subsidiary of Harcourt Brace Jovanovich, Publishers
New York   London
Paris San Diego San Francisco São Paulo Sydney Tokyo Toronto

Proceedings of
the Twelfth Annual
New York State Health Department
Birth Defects Symposium

ACADEMIC PRESS, INC.
111 Fifth Avenue, New York, New York 10003

United Kingdom Edition published by
ACADEMIC PRESS, INC. (LONDON) LTD.
24/28 Oval Road, London NW1 7DX

Library of Congress Cataloging in Publication Data
Main entry under title:

Clinical genetics.

(Birth Defects Institute symposia)
Includes index.
1. Medical genetics. 2. Genetic counseling.
3. Abnormalities, Human--Diagnosis, Intrauterine. date
I. Willey, Ann M. II. Carter, Thomas P. (Thomas Perry),
date. III. Kelly, Sally. IV. Series. [DNLM:
1. Hereditary diseases--Diagnosis. QZ 50 C641]
RB155.C573 1982   616'.042   82-11634
ISBN 0-12-751860-6

CONTENTS

CONTRIBUTORS AND PARTICIPANTS

Stylianos Antonarakis, Department of Pediatrics, Johns Hopkins University School of Medicine, Baltimore, Maryland

Robin M. Bannerman, Division of Medical/Human Genetics, State University of New York at Buffalo, Buffalo, New York

Barbara A. Bernhardt, Division of Medical Genetics, Johns Hopkins Hospital, Baltimore, Maryland

Corinne D. Boehm, Department of Pediatrics, Johns Hopkins University School of Medicine, Baltimore, Maryland

David J. H. Brock, Department of Human Genetics, University of Edinburgh, Edinburgh, Scotland

Thomas Carter, Birth Defects Institute, New York State Department of Health, Albany, New York

M. Michael Cohen, Jr., Faculties of Dentistry and Medicine, Dalhousie University, Halifax, Nova Scotia, Canada

Philip K. Cross, Birth Defects Institute, New York State Department of Health, Albany, New York

Roy A. Gravel, Department of Genetics, Research Institute, Hospital for Sick Children, Toronto, Ontario, Canada

Linda C. Higgs, New York Medical College and Medical Genetics Unit, Westchester County Medical Center, Valhalla, New York

Kurt Hirschhorn, The Mount Sinai Medical Center, New York, New York

Ernest B. Hook, Birth Defects Institute, New York State Department of Health, Albany, New York

Lillian Y. F. Hsu, Prenatal Diagnosis Laboratory of New York City and New York University School of Medicine, New York, New York

Haig H. Kazazian, Jr., Department of Pediatrics, Johns Hopkins University School of Medicine, Baltimore, Maryland

Sally Kelly, Birth Defects Institute, New York State Department of Health, Albany, New York

Murray D. Kuhr, Genetics Laboratory, Letchworth Village Developmental Center, Thiells, New York

Harvey L. Levy, Department of Neurology, Harvard Medical School, and IEM-PKU Program Children's Hospital Medical Center, Boston, Massachusetts

Evelyn Lilienthal, Genetics Laboratory, Letchworth Village Developmental Center, Thiells, New York

John A. Lowden, The Research Institute, The Hospital for Sick Children, Toronto, Ontario, Canada

Kathi Mesirow, School of Medicine, Wayne State University, Detroit, Michigan

Henry L. Nadler, School of Medicine, Wayne State University, Detroit, Michigan

Eileen R. Naughten, Butlin Fellow in Metabolic Medicine, Children's Hospital, Dublin, Ireland

Anton Novak, The Research Institute, The Hospital for Sick Children, Toronto, Ontario, Canada

John M. Opitz, Shodair Children's Hospital, Helena, Montana

Theresa E. Perlis, Prenatal Diagnosis Laboratory of New York City, New York, New York

John A. Phillips III, Department of Pediatrics, Johns Hopkins University School of Medicine, Baltimore, Maryland

Ian H. Porter, Birth Defects Institute, New York State Department of Health, Albany, New York

John D. Rainer, New York State Psychiatric Institute and Columbia University, New York, New York

Phyllis Rembelski, School of Medicine, Wayne State University, Detroit, Michigan

Mary J. Seller, Paediatric Research Unit, Guy's Hospital Medical School, London, England

Lawrence R. Shapiro, Departments of Pediatrics and Pathology, New York Medical College and Medical Genetics Unit, Westchester County Medical Center, Valhalla, New York, and Genetics Laboratory, Letchworth Village Developmental Center, Thiells, New York

Marie-Anne Skomorowski, The Research Institute, The Hospital for Sick Children, Toronto, Ontario

Paula M. Strasberg, The Research Institute, The Hospital for Sick Children, Toronto, Ontario, Canada

Dorothy Warburton, Departments of Human Genetics and Development, and Pediatrics, College of Physicians and Surgeons, Columbia University, New York, New York

Ann M. Willey, Birth Defects Institute, New York State Department of Health, Albany, New York

Patrick L. Wilmot, Department of Pathology, New York Medical College and Medical Genetics Unit, Westchester County Medical Center, Valhalla, New York

Gregory B. Wilson, Departments of Basic and Clinical Immunology and Microbiology, and Deparment of Pediatrics, Medical University of South Carolina, Charleston, South Carolina

PREFACE

In the practice of genetic diagnosis and counseling, there are a number of relatively common problems that we as geneticists and/or clinicians share. It was around just such concerns that this XII Birth Defects Symposium was organized. Each presentation was intended to provide the latest available information, a look at current methodologies, a description of current problems or limitations, and a preview or guide for the future coupled with practically applicable material. It is notable that the following chapters meet these criteria.

Some problems arise because of the speed of development and complexity of new methodologies. The recent, vast advances in the area of molecular genetics with recombinant DNA technology put us on the edge of major breakthroughs in the management of genetic diagnosis and discussion of those results with our patients, i.e., genetic counseling. Dr. Kazazian's description of restriction enzyme site detection and prenatal diagnosis of hemoglobinopathy certainly brings us to the perceived brink of such current technology. Mass application, however, may remain a problem.

Other problems arise because of lack of adequate information, as with counseling for mental retardation of unknown etiology, for idiopathic dysmorphic syndromes, and for psychiatric disorders. Concurrently, the growing store of information in other areas may lead to its own problems, as in the interpretation of prenatal cytogenetic diagnosis.

The discussion by Dr. Seller of preconceptual vitamin supplementation as a possible means of reducing recurrence risks for NTD brings great optimism into the field. However, in this same volume the four presentations devoted to cystic fibrosis may best represent the state of diagnosis and counseling in so many conditions. For all of the massive effort in laboratory studies, we are left without identification of the primary gene and therefore cut off from application of the new molecular technologies. Nevertheless, the desperate attempt to develop carrier detection assays and prenatal diagnostic methods continues. In the end we are left with a medically significant genetic disease for which counseling must still consist largely of an explanation of risk and a limited discussion of reproductive options.

We believe that this volume contains a great quantity of practical information applicable to counseling situations for selected diagnoses, as well as serving as a summary of the current limitations of diagnosis and counseling for genetic disorders.

ACKNOWLEDGMENTS

The annual Birth Defects Symposium, which gives rise to these volumes, owes much of its success to the administrative expertise of Luba Goldin and her assistant, Kathy Ruth. The entire clerical staff is also to be thanked for their eager participation in staffing the many administrative positions. Additionally, the technical presentation of the speakers' audio/visual aides was dependent on the excellent help of Phil Cross, Norma Hatcher, and Nanette Healy.

The preparation of this text was the able production of Dorothy Fischer. Her attention to detail, with an artistic eye, was invaluable. Kathy Miller should also be thanked for her patience in proofreading. We also thank Dr. David Axelrod, Commissioner of Health, Dr. Glenn Haughie, Director of Public Health, and Dr. David Carpenter, Director of the Division of Laboratories and Research, for encouragement and support.

NUTRITIONAL SUPPLEMENTATION AND PREVENTION OF NEURAL TUBE DEFECTS

Mary J. Seller

In 1976, Smithells, Sheppard and Schorah[1] published a prospective study of the blood vitamin status of an unselected group of over 900 women in the first trimester of pregnancy. Their first finding was that there was a significant social class gradient in blood vitamin levels. Social classes I and II* had significantly higher levels of red cell folate, leucocyte vitamin C, riboflavin and serum vitamin A, and higher (but not significantly so) serum folate, than classes III, IV and V. Their second finding was that the six women who subsequently gave birth to infants with neural tube defects (NTD) had, in the first trimester, lower serum and red cell folate, white blood cell vitamin C and riboflavin than women who had children without NTD. This finding held even when the social classes were taken into account, and despite small numbers the differences were significant for red cell folate and white blood cell vitamin C. Smithells and his colleagues submitted that these findings were compatible with the hypothesis that nutritional deficiencies are significant in the causation of neural tube defects.

As a consequence, they planned an intervention study which was subsequently undertaken in a collaborative way by five centers in the United Kingdom and started in 1976. This work, to be described, has been done jointly by myself at Guy's Hospital, London; by R.W. Smithells, S. Sheppard and C. Schorah at Leeds University; by N.C. Nevin, Queen's University, Belfast; by R. Harris and A. Read of Manchester University; and by D. Fielding, Chester Hospital, and has been published.[2,3] The incidence of NTD varies geographically within the United Kingdom, from around 2.9/1000 births in the

---

*Social groupings are made according to the occupation of the husband, as laid down by the Registrar General. They are correlated with such factors as education and economic environment, but have no direct relationship to the level of remuneration of particular occupations. The groupings are as follows:

I = Professional   IV = Partly skilled
II = Intermediate   V = Unskilled
III = Skilled occupations: non-manual (N), manual (M)

southeast to about 8/1000 births in Northern Ireland. Our study incorporated areas of low, high and medium incidence.

Women who had already had one or more children with NTD, who were contemplating another pregnancy, but were not yet pregnant, were offered periconceptional multivitamin supplementation. Originally a double-blind placebo trial was planned, and indeed, the placebo tablets were already prepared. However, this was rejected independently by both the Leeds and Guy's Hospitals' Ethical Committees, the first two centers to be involved in the project, and so a less scientifically satisfactory experimental design was adopted.

The women who agreed to participate all received vitamin tablets, Pregnavite Forte F (Bencard), which in a daily dose of three tablets provides a total of 4000 IU vitamin A, 400 IU vitamin D, 1.5 mg thiamine, 1.5 mg riboflavin, 1 mg pyridoxine, 15 mg nicotinamide, 40 mg ascorbic acid, 480 mg calcium phosphate, 0.36 mg folic acid and ferrous sulphate equivalent to 75.6 mg Fe. They were asked to take them for at least a month before conceiving and for the first eight weeks of their pregnancy. Those who complied with this regime form a "fully supplemented" group. The women actually took the vitamin tablets for varying lengths of time from the minimum 28 days to many months, according to how long it took them to achieve pregnancy; the mean was 110 days. During the supplementation period, tablets were sent regularly to the women by post at one to three month intervals.

Some women who had agreed to take part conceived before they had taken a full month's therapy; others were found to have begun supplementation a few days after conception. All these women were placed in a separate group which was called "partially supplemented". The women were assigned to these two groups when they reported to us that their pregnancies were confirmed at around seven to eight weeks' gestation. At this time the period over which they had been receiving supplementation was calculated.

In the absence of being able to do a properly controlled trial, the eventual "control" women were those who were at the same risk of producing NTD as the supplemented women, living in the same general area, who conceived within the same month, and as near as possible in age to the supplemented women. All but one center matched control and supplemented individuals one for one; Belfast was not able to do this. The control women were selected prospectively when they applied to our genetic counseling clinics at 8-12 weeks gestation because they wanted to book an amniocentesis. They comprised "at risk" women who did not participate in the trial because they were not previously known to us, or in a few cases were women who had been approached but had refused to participate. It should be noted that since the controls were picked at this early stage in pregnancy, long before maternal serum alpha fetoprotein screening or amniocentesis took place, no bias could be introduced by selecting those women who were especially likely to have a recurrence of NTD. The only known difference between the supplemented and control women is that early first trimester (before eight weeks) spontaneous abortions are re-

TABLE 1
Results of Periconceptional Multivitamin Supplementation
with Recurrences of NTD in Parenthesis

	Unsupple-mented	Supplemented Fully	Partially
Total Mothers	300 (13)	200 (1)	50
Add: twin pairs	5	4	2
Total Babies/Fetuses	305 (13)	204 (1)	52
Deduct: spontaneous abortions not examined	10	9	1
Total Babies/Fetuses Examined	295 (13)	195 (1)	51

corded for the former, but are necessarily absent in the latter because of the manner of selection of the controls after eight weeks from amniocentesis candidates.

The pregnancies of all women were followed. If amniocentesis was performed the result was recorded and also the eventual outcome of the pregnancy.

The results are shown in Table 1. They take into account twin pregnancies, and also the fact that a number of pregnancies ended in spontaneous abortion. Although every effort was made to examine the products, this was not possible in every case and in some which were examined there was no recognizable fetus.

There were 300 unsupplemented control women who had 305 babies or fetuses of which 295 were examined. There were 13 NTD, a recurrence of 4.4%. There were 200 fully supplemented women who had 204 babies or fetuses of which 195 were examined and there was one recurrence (0.5%). Among the 50 partially supplemented women (52 babies or fetuses, 51 examined) there were no recurrences. The difference between the fully supplemented and the control women is significant (p<0.01, one tailed, Fisher's exact method).

The most likely explanation for the results is that vitamin supplementation has somehow prevented NTD. However, it is possible that the results might have been unwittingly biased in some way, so the results are now examined for that possibility.

First, if the results are broken down into the different individual centers (Table 2), it can be seen that there is an excess of controls from Belfast, the high incidence area of the United Kingdom. This might have advantageously

TABLE 2

Number of Women Participating in Periconceptional Multivitamin
Supplementation According to Center with Recurrence of NTD in Parenthesis

	Unsupplemented	Supplemented Fully	Partially
Leeds	35 (2)	38	8
London	95 (5)	70 (1)	24
Belfast	124 (6)	53	9
Manchester	36	31	8
Chester	10	8	1
Total	300 (13)	200 (1)	50

TABLE 3

Results of Periconceptional Multivitamin Supplementation According to
Number of Previous Affected Children with Recurrence of NTD in Parenthesis

	Unsupplemented	Supplemented Fully	Partially
1 previous NTD	270 (9)	182 (1)	47
% of total	90%	91%	94%
2 previous NTD	30 (4)	18	3
% of total	10%	9%	6%
Total	300 (13)	200 (1)	50

biased the number of recurrences in the controls. But this is not so because the recurrences were 2 in 35 in Leeds (5.7%), 5 in 95 in London (5.3%) and 6 in 124 in Belfast (4.8%).

From this breakdown by center it can also be seen that the one recurrence in the supplemented group was in the southeast of England, the low incidence area of the United Kingdom. While this might be chance, it is possibly significant. It may be that the therapeutic effect of an environmental agent is less in areas where the incidence of NTD is lower.

Second, the recurrence of NTD is known to be higher after two previous affected children than after only one, and a bias might have arisen if there were more of these higher risk women in the controls. When women are separated according to their risk category (Table 3), there is no excess of the former in the control women. Nine percent of fully supplemented women, and 10% of control women had two previous affected offspring.

It has recently been shown[4] that not only is there a higher incidence of NTD, but also a higher recurrence among the lower socioeconomic classes. Analysis of the social class of our women (Table 4) shows that the proportion of women in the lower classes (conventionally regarded as IIIM, IV and V) was higher in the control group (64% of all the control women) than in the fully supplemented women (55% of the total). However, the large difference overall in the recurrence of NTD between the supplemented and control women cannot be accounted for by this relatively small excess of women of the lower social classes among the controls. Notwithstanding, vitamin therapy seems beneficial to this group. If the effects of vitamin therapy on the lower social class group of women alone are examined, the recurrence is 11 in 191 (5.7%) and 1 in 110 supplemented women (0.9%), a statistically significant difference ($p = 0.02 > 0.01$). For women of classes I, II and IIIN combined, the respective figures are 2 in 87 controls (2.2%) and 0 in 86 supplemented (0%).

Finally, it is possible that the lack of recurrence in the supplemented group might be accounted for by the fact that fetuses with NTD are being spontaneously aborted. However, the incidence of spontaneous abortion in the supplemented and control groups is similar (Table 5), being 10% and 9.6% respectively. Further, of a total of 13 abortuses examined from supplemented women, none had NTD, while one of 19 from control women had an NTD.

TABLE 4
Results of Periconceptional Multivitamin Supplementation
According to Social Class

| | Classes | | | Classes | | | | | |
| | I | II | IIIN | IIIM | IV | V | Unknown | | |
	US	FS	PS	US	FS	PS	US	FS	PS
Number	87	86	23	191	110	23	22	4	4
% of total	29%	43%	46%	64%	55%	46%	7%	2%	8%
Recurrence	2			11	1				

US = unsupplemented, FS = fully supplemented, PS = partially supplemented

TABLE 5
Numbers of Spontaneous Abortions with Recurrence of NTD in Parenthesis

Spontaneous Abortions	Unsupplemented	Fully Supplemented
Examined	19 (1)	11
Not examined	10	9
Total spont. abortions	29 = 9.6%	20 = 10%
Total women	300	200

We have also considered differences in maternal age and previous reproductive history which likewise seem negative. Thus, there is nothing we can
think of to explain our results other than that vitamin supplementation has, in
some way, prevented the recurrence of NTD, but we acknowledge that we have
not yet proven this.

We are currently undertaking a second, identical trial, and the nearly completed results show that it is going exactly the same way.

The lack of a controlled double-blind placebo trial has meant that we
cannot be sure that a special group of low-risk women has not selected itself
for vitamin therapy. Further, because of the "cocktail" nature of the vitamin
preparation we used, even if the therapy really is effective, it is not known
which vitamin or vitamins are beneficial. Consequently, further studies in the
field are to be encouraged.

In this connection, a related clinical trial was published earlier this year
which has evoked much interest. It was a periconceptional folate supplementation study by Laurence, James, Miller, Tennant and Campbell[5] in
South Wales. I have been asked to examine critically this study. It is vital
this is done because this paper entitled "Double-blind randomized controlled
trial of folate treatment before conception to prevent recurrence of neural
tube defects" states in its summary, "There were no recurrences among those
who received supplementation and six among those who did not, this difference is significant (p = 0.04)." This statement is of enormous potential importance. The results as stated appear compelling and the study seems to be
the good scientific trial that our own failed to be. Also, it specifically used
a single substance where we used a compound of many, thereby isolating
which component could be the true therapeutic agent.

However, although folate may well be the definitive agent, it is not proven
by this study. I will go through the work of Laurence and his colleagues stage
by stage.

Women who had a previous child with a neural tube defect were traced and visited, and told of the proposed study. Those who agreed to participate had their dietary history recorded and a blood sample taken for serum and red cell folic acid assay. They were asked to take a tablet containing either 2 mg folic acid or a placebo twice a day starting from the time contraceptive measures were stopped. Women were assigned to folate treatment or placebo by random numbers and neither they nor Laurence and his colleagues knew the content of the tablets. Thus a double-blind, randomized controlled trial was truly set up.

The women reported to Laurence and his colleagues when their pregnancy was first suspected and were revisited as soon as possible thereafter. At this visit their diet and general health were noted and, although not explicitly stated in the method in the paper, they were asked whether they had actually taken the tablets allocated to them. Those who acknowledged that they had not were immediately excluded from the study and there is no record of these women. The women who said they had taken the tablets had a second blood sample taken for folate estimation, and they were revisited at six months and after delivery.

Of over 300 women who were randomized into the trial to receive folate or placebo tablets, there were only 111 women who achieved a pregnancy during the trial period and who satisfied the clinical criteria, that is, when interviewed in the first trimester they said they had taken the tablets and recognized their pregnancy during the first seven weeks following conception. When the code was broken 60 of these 111 women were found to have received folate supplementation and 51 the placebo. Two of the 60 supplemented women had a recurrence of an NTD, and 4 of the 51 placebo women had another NTD child; there is no significant difference between the two groups. It must be emphasized that this is the result of the double-blind, randomized controlled trial which was unsuccessful.

The study was finished and the code was broken in 1975. At that stage, as planned at the beginning of the study, the 60 women who had been allocated folate tablets , and who said they had taken them, were separated into "compliers" and "noncompliers" regarding tablet taking by Laurence and his colleagues on the basis of their serum folate levels in the first trimester, using a cut-off point of 10 μg/1. Thus, despite what the women themselves said, Laurence et al judged whether they had taken their tablets or not according to their serum folate levels.

Using this criterion, 45 of the 60 women were designated "compliers" and 15 "noncompliers". One of the recurrences of NTD occurred in the former and one in the latter. However, the patient who had the recurrence in the compliance group, whose folate level was enormously elevated (212 μg/1), was subsequently transferred to the noncompliance group (definition: serum folate level less than 10 μg/1) because on requestioning in 1980 she admitted that she had not taken the folate tablets during early pregnancy but had taken a large number the day before the field worker visited her. With this information, this woman should actually have been completely excluded along with all the other women who admitted they had not taken the tablets in the first trimester, Instead, she was transferred to the "control" group.

So, from the group of 60 folate supplemented women, Laurence *et al* were able to end up with 44 women whom they designated "compliers" on the basis of their serum folate levels, among whom there were no recurrences. There were two recurrences among the 16 women whom they called "noncompliers". The placebo group was unchanged because compliance could not be tested; there were 51 women with 4 recurrences. Serum folate levels are given for women in this group with recurrences of NTD. It is interesting to note that one had a level of 11 μg/1, showing that the cut-off point of 10 μg/1 used to divide up the folate women does not therefore differentiate, without overlap, folate takers and nontakers.

In the final analysis, it is stated that a significant difference (Fisher's exact test, p = 0.04) was found "between those who received supplementation and those who did not". It should be noted that while the former comprised the 44 women who had received folate tablets (who said they had taken them, and whose serum folate levels were above 10 μg/1), the latter (those who did not receive supplementation) is a mixed group. The mixed study group was composed of the 51 placebo treated women (four recurrences), combined with 15 women who had received folate tablets and who when interviewed in the first trimester said they had taken the tablets, but who were classified as noncompliers because their serum folate levels were below 10 μg/1 (one recurrence). The single woman who showed excessively high serum folate levels, but had not regularly taken the tablets unlike all the others who admitted they had not taken their tablets, was not excluded in this group (one recurrence). Thus, it is not the results of the double-blind trial which give the statistically significant results, but a comparison of recurrences in these regroupings of the women involved in the trial after the code was broken. The authors themselves point this out in the discussion, calling the significant findings the biological effects of folate treatment.

In an earlier paper[6] where this study is also reported, there is a third group of women included who "elected to take neither" (the folate or placebo). There was only one recurrence in this unsupplemented group of 54 women. This emphasizes how very different results can be obtained with different samples in trials where small numbers are involved. It is interesting to note that from this sample it might even be inferred that not taking folate is beneficial with regard to the recurrence of NTD.

Therefore, the study would seem to say that folate is effective in the prevention of NTD, but this is not yet proven. In Laurence's work, much hinges on the meaning of low serum folate levels. Such levels do not necessarily mean that the women did not take their tablets. The women with low levels might have a poor uptake of folate and this could be the key to their susceptibility to produce children with NTD. It is possible that the crux of the matter may reside in the folate status of the mother. Perhaps the most significant finding of the work may be that the women who had good blood levels of folate in the first trimester had normal children, while the recurrences of NTD occurred among those with lower levels. In this sense it agrees with the earlier observations of Smithells et al[1] in their first trimester blood vitamin study. The work also highlights another very important fact not discussed here, which is how vital it is to have an adequate diet.

In conclusion, the multivitamin trial of my colleagues and myself has produced significant results in reducing the number of recurrences of NTD in a cohort of 200 women, and a second cohort now nearly completed seems to be repeating this success. While not absolutely ideal scientifically, this approach has for the first time apparently succeeded in the primary prevention of a severe developmental defect. At present, no other comparable studies have been published, but it is known that several are currently being planned. The results of these are awaited with keen interest.

REFERENCES

1. Smithells RW, Sheppard S, Schorah CJ. Vitamin deficiencies and neural tube defects. *Arch Dis Childh* 1976; 51: 944-50.
2. Smithells RW, Sheppard S, Schorah CJ, et al. Possible prevention of neural tube defects by periconceptional vitamin supplementation. *Lancet* 1980; i: 339-40.
3. Smithells RW, Sheppard S, Schorah CJ, et al. Apparent prevention of neural tube defects by periconceptional vitamin supplementation. *Arch Dis Childh* 1981; 56: 911-18.
4. Nevin NC. Recurrence risk of neural tube defects. *Lancet* 1980; i: 1301-2.
5. Laurence KM, James N, Miller MH, Tennant GB, Campbell H. Double-blind randomized controlled trial of folate treatment before conception to prevent recurrence of neural tube defects. *Brit Med J* 1981; 282: 1509-11.
6. James N, Laurence KM, Miller M. Diet as a factor in aetiology of neural tube defects. *Z Kinderchir Grenzgeb* 1980; 31: 302-07.

ADDENDUM

The interim results of the second cohort are now published* and they repeat the success of the first study. In the unsupplemented group there were 10 recurrences of NTD amongst 198 pregnancies, while in 202 fully supplemented women there were two recurrences. If these results are added to those of the first cohort, the difference between the vitamin supplemented mothers (three recurrences amonst 397 offspring) and unsupplemented women (23 recurrences among 493 offspring) is highly significant ($p < 0.0003$, Fisher's exact test, one tailed).

*Smithells RW, Sheppard S, Schorah CJ, et al. Vitamin supplementation and neural tube defects. *Lancet* 1981; ii: 1425.

PRENATAL DIAGNOSIS OF HEMOGLOBINOPATHIES
BY RESTRICTION ANALYSIS: METHODOLOGY AND EXPERIENCE

Corinne D. Boehm
John A. Phillips III
Stylianos Antonarakis
Haig H. Kazazian, Jr.

The method of choice for prenatal diagnosis of hemoglobinopathies has been in a period of change for the past three years. Prior to 1978, prenatal diagnosis required a sample of fetal blood which was obtained by fetoscopy. By examining globin chain synthesis in fetal red cells, every pregnancy at risk for sickle cell anemia or β-thalassemia is potentially diagnosable. However, because of the specialized skill that is needed for fetoscopy, the procedure is available only at a very few centers. Due to this limited availability and a continuing accompanying 4–5% risk of fetal mortality, many couples at risk for a child with a hemoglobinopathy did not truly have the option of prenatal diagnosis available to them prior to 1978. In 1978 it became possible to detect sickle cell anemia prenatally in the majority of pregnancies at risk (62%)[1] by taking advantage of recently developed techniques in molecular genetics. While the percentage of potentially diagnosable pregnancies dropped from 100% by fetoscopy to 62% by this method, the actual number of pregnancies undergoing prenatal testing for sickle cell anemia rose because fetal samples are obtained by the much more widely available and safer technique of amniocentesis (fetal risk <.5%). Modifications to this second test have increased the applicability today to 95% of pregnancies at risk for sickle cell anemia.[2,3] For these tests, in use since 1978, DNA markers near the β globin gene are analyzed. Not all β globin genes have informative DNA markers; when they do not, prenatal diagnosis is not possible by this method. However, 95% of people with sickle trait do have marker DNA that can be used for these purposes. Therefore, in 90% of couples at risk for a child with sickle cell anemia, both members of the couple will have informative DNA markers and prenatal diagnoses can be obtained in every pregnancy. In the other 10%, one member of the couple will have marker DNA. For them, sickle cell anemia can be ruled out in 50% of their pregnancies. In total, 95% of pregnancies at risk can be diagnosed prenatally. There is currently another prenatal test being developed.[4] This

CLINICAL GENETICS: PROBLEMS IN
DIAGNOSIS AND COUNSELING

11

test examines DNA at the actual point of mutation of the β^A globin gene to the β^S gene. Because each sickle gene has the same mutation, it is possible to predict fetal β globin genotype in every pregnancy by this newest technique. This test will be the method of choice for prenatal diagnosis when it is completely developed. Presently, for couples at risk for β-thalassemia, the DNA methods are useful in 85% of pregnancies when family studies are carried out.[5]

However, I'll now direct attention to the prenatal test for sickle cell anemia that makes use of marker DNA near the β globin gene. Since 1978 we have studied 40 families at risk for a child with sickle cell anemia using this method.

The cause of sickle cell anemia is an amino acid substitution in the β globin protein. This variant protein results from a single base (nucleotide) change in the β globin gene. The informative markers in the DNA are located right next to the β globin gene. Because of the proximity of the markers to the β gene they are essentially always inherited together. The markers are used to "represent" the β globin genes. Just as a person with sickle cell trait has two different β globin alleles (the normal β^A and variant β^S), so he must also have two different forms of DNA at a marker site if we intend to distinguish between his two different β globin alleles by use of the marker DNA. The markers are, in a sense, independent of the β^A and β^S globin alleles. We are just pressing them into service for our own purposes. The important point for prenatal diagnosis is that a trait parent must have two different forms of DNA at a marker site.

These representatives of the β globin gene are known as polymorphic DNA restriction sites (also known as restriction fragment length polymorphisms). These DNA sites are polymorphic (have more than one form) with respect to their ability to be cleaved by specific restriction endonucleases. Each restriction endonuclease, and there are about two hundred of them, is an enzyme that cleaves DNA non-randomly by recognizing specific nucleotide sequences.

Figure 1 shows a map of the β globin region. Depicted are the $^G\gamma$, $^A\gamma$, δ, and β globin genes in their proper orientation along chromosome 11. Also sketched in are so-called pseudo β (ψ β) genes. The nucleotide sequences of these pseudo genes are similar to the β globin gene, yet defects in their structure preclude their expression into protein through lack of appropriate transcription signals. Also shown in Figure 1 are the locations of eight well-established polymorphic sites for the enzymes *Hind* III, *Hpa* I, *Bam* HI, *Hinc* II, and *Ava* II, each with its own recognition sequence.[1,3,6-8] This whole stretch of DNA is sufficiently short (60,000 base pairs, or 60 kb) that the region is essentially devoid of recombination with each meiosis, meaning that the region almost always gets inherited as a unit.[9] In some chromosomes, *Hpa* I will cleave the DNA at a point 7 kb downstream, or 3', to the β gene; in other chromosomes it will not. This ability to be cleaved or not is constant for a particular chromosome and is also inherited, so that each descendent of a chromosome will retain its parent's ability to be cleaved or not. All along

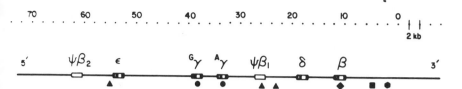

Fig. 1. Linkage map of the β globin gene cluster[10] with the location of eight high-frequency polymorphic restriction sites. ▲ = *Hinc* II. ● = *Hind* III. ♦ = *Ava* II. ■ = *Hpa* I. ◖ = *Bam* HI.

HPA 1 RESTRICTION MAP NEAR β-GLOBIN LOCUS

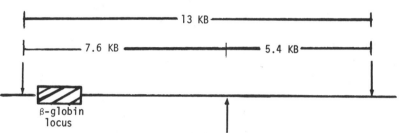

Fig. 2. Constant and polymorphic *Hpa* I restriction sites flanking the β globin locus. KB = kilobase. ↓ = constant cleavage site. ↑ = polymorphic cleavage site.

RESTRICTION ENDONUCLEASE ANALYSIS

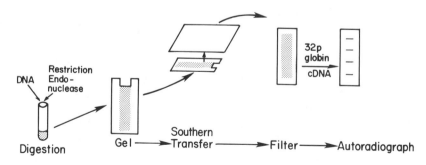

Fig. 3. Schematic representation of the steps in restriction endonuclease analysis.

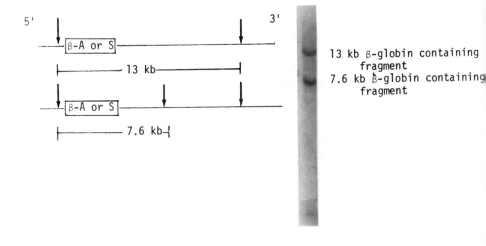

Fig. 4. Heterozygosity at polymorphic *Hpa* I site depicted (a) schematically and (b) on autoradiogram.

the chromosome are many more cleavage sites which are not polymorphic and, therefore, are always present. A more detailed map of both constant and polymorphic *Hpa* I sites flanking the β globin locus is shown in Figure 2. Chromosomes which have this 3' *Hpa* I polymorphic cleavage site will produce a 7.6 kb fragment of β globin containing DNA after being treated with enzyme *Hpa* I; chromosomes lacking this cleavage site will produce a 13 kb fragment. This presence of the polymorphism can be detected by altered sizes of DNA fragments after digestion with that particular enzyme (Figure 3). The cleavage process by the enzyme is known as digestion of the DNA. After digestion with the enzyme, the DNA fragments are electrophoresed in a gel that separates them by size. DNA is normally double-stranded. While still in the gel, the DNA is treated with sodium hydroxide (NaOH), which breaks the hydrogen (H) bonds holding the two strands together. The fragments, in single-stranded state, are transferred out of the gel and onto a nitrocellulose filter which is then exposed to radioactive single-stranded copies of the gene contained in the fragments of interest. The DNA on the filter, because it is single-stranded, is available to resume the native, or double-stranded, conformation by re-establishing H bonds with single-stranded DNA of the same or nearly same nucleotide sequences. After hybridization with radioactive sequences of the specific DNA, the fragment sizes of interest are detectable by exposing the filter to X-ray film (a process called autoradiography).

Figure 4 shows the *Hpa* I map from a person whose two chromosomes are different with respect to the polymorphic *Hpa* I site; one has the *Hpa* I cleavage site, and one lacks it. This difference is detectable in the accompanying

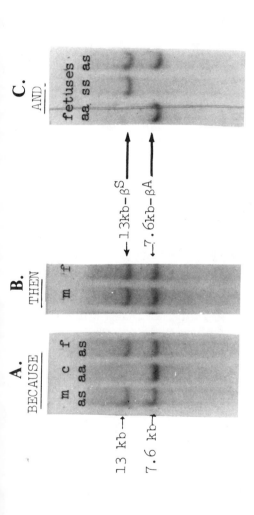

Fig. 5. (A) *Hpa* I restriction patterns from a trait mother (m/as), trait father (f/as), and their homozygous normal child (c/aa).
(B) Coupling phases between parental β globin alleles and polymorphic *Hpa* I restriction sites.
(C) *Hpa* I restriction patterns predicting indicated fetal β globin genotypes.

autoradiogram. Electrophoresis has been carried out from top to bottom, with the smaller fragments migrating more quickly than the larger. Because the polymorphic sites are functionally independent of the β globin gene, in one trait individual the 13 kb fragment may represent the β^S allele and the 7.6 kb fragment the β^A allele, while in another trait individual the opposite could be true with the 13 kb fragment representing the β^A allele and the 7.6 kb fragment the β^S allele. The orientation of these sites in a particular person will be inherited. Establishing the correct orientation in a sickle trait individual with two fragment sizes is essential for prenatal diagnosis because we predict which β globin alleles the fetus has inherited by examining the fetal fragment sizes. The orientation, or coupling phases, is determined by following the inheritance patterns of both the β globin alleles and the fragment lengths in a particular family (Figure 5). As shown in Figure 5, if two trait parents have both the 13 kb and the 7.6 kb *Hpa* I fragments and their normal, non-trait child has only the 7.6 kb fragment, we then know that both parents' β^S alleles reside in their 13 kb fragments. Consequently, when we observe which fragments the fetus has inherited from this couple, we can predict which β globin alleles it has inherited.

For at-risk couples who do not have any children, coupling phases can be determined by use of the couples' parents. When there is no family available for study, we can usually predict by statistical methods which fragments will contain each β globin allele. This is possible because of what is known as linkage disequilibrium. Only 8% of β^A alleles reside in 13 kb *Hpa* I fragments, whereas 65% of β^S alleles do. Because of these differences, 95% of people who have sickle cell trait and both the 13 and 7.6 kb *Hpa* I fragments will have their β^A allele contained in the 7.6 kb fragment and their β^S allele in the 13 kb fragment. Results of fetal *Hpa* I fragment lengths will then indicate which β globin alleles the fetus has inherited. However, the accuracy of a prediction in such a case is a direct result of how certain we are of the parents' *Hpa* I/β globin orientations after a statistical analysis and depends on which polymorphic sites are being used. This statistical analysis is used only when absolutely no other family members are available through whom we can follow inheritance patterns and therefore be 100% certain of the correct orientations.

In the event that only one parent has informative markers at any one of the polymorphic sites, there is a 50% chance with each pregnancy that analysis of fetal DNA might rule out an affected fetus (Figure 6). In this example, only the mother has useful marker DNA. Because her SS child inherited both her β^S allele and her *Hpa* I 13 kb fragment, we know the two are coupled, or on the same chromosome. Fetuses inheriting the mother's *Hpa* I 7.6 kb fragment would be either AS or AA, depending on the paternal contribution, while those inheriting her *Hpa* I 13 kb fragment would be either AS or SS. The father's contribution cannot be detected by these methods. In the 5% of pregnancies in which SS cannot be excluded by this means or all other

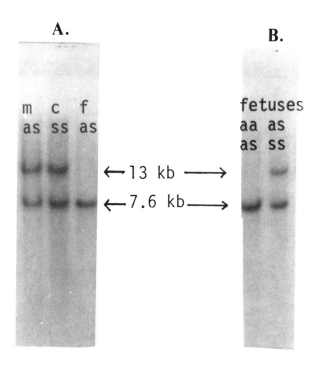

Fig. 6. (A) *Hpa* I restriction patterns from a trait mother (m/as), trait father (f/as) and affected child (c/ss).

(B) *Hpa* I restriction patterns predicting indicated fetal β genotypes.

available markers, further delineation of fetal β globin genotype would require fetoscopy.

When we receive blood (our source of DNA from people) from a family who will desire prenatal diagnosis in the future, we begin a search for markers with which we can distinguish the β globin alleles. Although there are eight sites that can be examined, not all prove to be equally useful for this purpose. Sixty-three percent of trait individuals have different *Hpa* I cleavage patterns on their two chromosomes. This was the only site that was used with the initial tests done in 1978. Using both *Hpa* I and *Hind* III sites, 93% of trait individuals will have different patterns. Using all sites, 95% of trait individuals are heterozygous (or have different cleavage patterns) at one or more of these sites. Similarly, using all sites in couples at risk for β-thalassemias, about 82% of trait individuals are heterozygous at one or more sites. Once we have established which sites are useful in a particular family and what the orientation of the sites and β alleles are through family studies, we are ready to analyze a sample of fetal DNA. There is a significant amount of work that goes into the initial stages of this type of analysis, and thus it is best for the family at risk to begin the study as early as possible before the 16th week of pregnancy when amniocentesis should be performed. In this way families know what their chances are of obtaining a complete prenatal diagnosis and how long it might take, referring counselors and doctors know whether they might need to schedule a back-up fetoscopy, and we know with which enzymes we should digest the fetal sample. From the time we obtain blood samples, it takes two to three weeks to analyze the various sites and determine the applicability of this procedure for a particular family. The time of analysis for the fetal sample varies and can take anywhere from two to five weeks, depending on whether one or two digestions are needed for complete analysis. Two digestions are needed in the event that one parent has a marker at one site (such as the *Hpa* I site) and the other has a marker at another site (such as a *Hind* III site).

Of the 40 families at risk for sickle cell anemia we have studied, 30 have actually undergone prenatal testing. The remaining ten are either not yet pregnant, have decided against prenatal testing, or pregnancy was terminated either spontaneously or by choice prior to study. The 30 prenatal diagnoses made are not significantly different from what would be expected from an autosomal recessive inheritance pattern: 18% have been AA, 48% AS, and 33% SS. More specifically, the breakdown of predictions is 3 AA, 11 AS, 9 SS, 5 AA or AS (sickle cell disease ruled out), and 2 AS or SS (disease not ruled out). Thirty-three percent of the predictions have been confirmed; the remainder are either awaiting birth for confirmation, lost to follow-up through the referring center, or, as is frequently the case in affected and terminated pregnancies, impossible to confirm after the termination process. Of the ten pregnancies bearing an affected fetus, eight were terminated. There were an additional two cases, not included in the 30 families, in which the women themselves had sickle cell anemia and their spouses had sickle cell trait. Both fetuses were predicted as affected, and neither pregnancy was terminated. To date, no diagnostic errors have been made to our knowledge.

We have also applied similar tests to the prenatal diagnosis of the following hemoglobinopathies: β-thalassemia 25, S/C disease 4, S/β^{thal} 3, α-thalassemia 3, β-thal/Lepore 1, and S/O^{Arab} 1. Prenatal testing for β-thalassemia by amniocentesis alone has been possible in 11 of 15 couples at risk whose testing is complete, while 4 of 15 couples also required fetoscopy. Since β-thalassemia mutations usually do not alter restriction endonuclease sites or produce new ones, nearly all prenatal diagnoses of β-thalassemia by DNA methods will continue to require linkage analysis of markers other than the mutations themselves.

In contrast, the newest test for prenatal diagnosis of sickle cell anemia detects the sickle mutation itself and is much less complicated than the one described above. The nucleotide change producing the sickle cell gene obliterates a recognition sequence for the enzyme *Dde* I that is normally found in β^A genes. The altered restriction pattern is detectable by electrophoretic and hybridization techniques similar to those previously described. When this test becomes available in early 1982, it will (a) work for all pregnancies at risk for sickle cell anemia and (b) do away with the need for the time-consuming and complicated tasks of searching for polymorphic sites and establishing correct coupling phases by family studies.

ACKNOWLEDGMENT

This work was supported by grants from the March of Dimes Birth Defects Foundation (6-194) and the National Institutes of Health (AM 13983).

This work was presented at the Postgraduate Conference "The Sickle Cell Spectrum: A family of related syndromes". The proceedings of the conference will be published by Alan R. Liss, Inc., and a modified form of this manuscript will appear in that publication.

REFERENCES

1. Kan YW, Dozy AM. Polymorphism of DNA sequence adjacent to human β globin structural gene: Relationship to sickle mutation. *Proc Natl Acad Sci* 1978; 75: 5631.

2. Phillips JA, Panny SR, Kazazian HH Jr., Boehm CD, Scott AF, Smith KD. Prenatal diagnosis of sickle cell anemia by restriction endonuclease analysis: *Hind* III polymorphisms in γ globin genes extend test applicability. *Proc Natl Acad Sci* 1980; 77: 2853.

3. Antonarakis SE, Boehm CS, Giardina PJV, Kazazian HH Jr. Non-random association of polymorphic restriction sites in the β gene cluster. *Proc Natl Acad Sci*. 1982; 79: 137.

4. Geever RF, Wilson LB, Nallaseth FS, Milner PF, Bittner M, Wilson JT. Direct identification of sickle cell anemia (SS) by blot hybridization. *Proc Natl Acad Sci.* 1981; 78: 5081.
5. Phillips JA III, Vik TA, Scott AF, *et al.* Unequal crossing-over: A common basis of single α globin genes in Asians and American Blacks with hemoglobin-H disease. *Blood* 1980; 55: 1066.
6. Jeffreys AJ. DNA sequence variants in the $^G\gamma$, $^A\gamma$, δ, and β globin genes of man. *Cell* 1979; 18: 1.
7. Tuan D, Biro PA, deRiel JK, Lazarus H, Forget BG. Restriction endonuclease mapping of the human γ gene loci. *Nucl Acid Res* 1979; 6: 2519.
8. Kan YW, Lee KY, Furbetta M. Polymorphism of DNA sequence in the β globin gene region: Application to prenatal diagnosis of β^O-thalassemia in Sardinia. *New Engl J Med* 1980; 302: 185.
9. Kurnit DM, Hoehn H. Prenatal diagnosis of human genome variation. *Ann Rev Genet* 1979; 13: 235.
10. Efstratiadis A, Posakony JW, Maniatis T, *et al.* The structure and evolution of the β globin gene family. *Cell* 1980; 21: 653.

ADDENDUM

In late 1981, a very good direct prenatal test for sickle cell anemia was developed using the restriction enzyme, Mst II. The β^S mutation alters an Mst II site giving rise to a 1.4 kb fragment from β^S genes. A 1.2 kb fragment is generated from β^A genes. All couples at risk are now eligible for prenatal testing without linkage analysis, and we have used the test successfully in seven of seven couples tested. (Orkin SH, Little PFR, Kazazian HH Jr., and Boehm CD. *New Engl J Med.* In press.)

CURRENT CONCEPTS OF TREATMENT IN PHENYLKETONURIA

Eileen R. Naughten
Harvey L. Levy

When Sir Archibald Garrod delivered his famous Croonian Lectures before the Royal College of Physicians in 1908,[1] he said those inborn errors of metabolism "would attract earliest attention which advertise their presence in some conspicuous way, either by some strikingly unusual appearance of surface tissues or of excreta, by the excretion of some substance which responds to a test habitually applied in the routine of clinical work, or by giving rise to obvious morbid symptoms. Each of the known inborn errors of metabolism manifests itself in one or other of these ways and this suggests that they are merely the most obvious members of a far larger group, and that not a few other abnormalities which do not so advertise their presence may well have escaped notice hitherto."

Phenylketonuria (PKU) does not advertise its presence in a conspicuous manner and therefore was not discovered until the mid-1930s. However, we now know that when PKU was identified by Fölling,[2] it fulfilled the prophesy of Garrod. There is a far larger group of inborn errors than perhaps Garrod anticipated.

PKU continues to be among the most exciting of the inborn errors, despite (or because of) the large amount of study and attention it has received during the past 20 years. It is still leading us into new biochemical pathways, new concepts of pathogenesis of brain damage, and interest in new forms of treatment. We now have a much greater understanding of PKU. However, many old questions remain unanswered and several new ones have arisen. The number of these questions dramatically increased after Bickel's discovery of dietary treatment for PKU in the mid-1950s[3] and Guthrie's development of a newborn screening test for PKU in the early 1960s[4] (Table 1).

NEWBORN SCREENING

A new era began with routine newborn screening. Before this, treatment for PKU commenced after the patient had already presented with mental retardation. Now babies with PKU are identified on the basis of a biochemical finding and before signs of brain damage appear. Specific dietary therapy is

CLINICAL GENETICS: PROBLEMS IN
DIAGNOSIS AND COUNSELING

TABLE 1
Milestones in Phenylketonuria

Event	Major Figure	Year	Ref.
Clinical discovery	Fölling	1934	2
Determination of genetic pattern	Jervis	1939	5
Hyperphenylalaninemia	Jervis	1940	6
Enzyme defect	Jervis	1953	7
Dietary therapy	Bickel	1953	3
Newborn screening	Guthrie	1961	4
Mass screening	MacCready	1963	8

started which is aimed at minimizing the biochemical abnormality and thus preventing the brain damage.

Also, screening has led to information about the incidence of PKU worldwide. For instance, in Ireland the incidence is 1:6,000 (Cahalane SF. Personal communication, 1980.), whereas in Japan the incidence is only 1:60,000 (Naruse H, Kitagawa T, Oura T, et al. Personal communication, 1980.). In the United States, the PKU incidence is generally about 1:14,000.[11] PKU is unusual in some ethnic groups and almost unknown among blacks and Ashkenazi Jews. In certain large countries, such as China, the occurrence rate remains unknown because newborn screening is not conducted in these areas.

DIETARY TREATMENT

I. Criteria for Diet

Follow-up studies have established that PKU as initially detected by newborn screening is a group of conditions, each of which is biochemically and genetically distinct but which presents with an elevated blood phenylalanine (phe) level. Listed in Table 2 are the recognized metabolic disorders that result in hyperphenylalaninemia, their usual range of blood phe concentrations, and the established enzyme deficiency.

Most infants with phenylalanine hydroxylase deficiency whose blood phe level exceeds 12 or 15 mg/dl will require dietary treatment. Infants with defects in pterin metabolism, specifically dihydropteridine reductase deficiency and a defect in biopterin synthesis (as yet unspecified), require specific non-dietary therapy (see later section).

II. The Low PHE Diet

Low protein. The object of the diet for PKU is to lower the blood phe level by reducing the intake of this essential amino acid. One mechanism for accomplishing this is to reduce the amount of protein ingested. However, the degree of dietary protein restriction that is necessary to achieve this degree of phe restriction results in protein malnutrition unless the diet is supplemented by a special preparation that contains protein equivalents but little or no phe. (see below).

TABLE 2
Metabolic Disorders that Result in an Increased Blood Phenylalanine Level

Disorder	Blood phe* (mg/dl)	Enzyme deficiency
PKU	>20	PH <1%**
Atypical PKU	>12–20	PH 2–3%
Mild hyperphenylalaninemia	>2–12	PH 2–20%
Transient hyperphenylalaninemia	>2–20	?
DHPR† deficiency	12–20	DHPR 1%
Biopterin synthetase deficiency	12–20	?

*Blood phenylalanine values in corresponding SI units are 1200 umoles/L (20 mg/dl); 720 umoles/L (12 mg/dl): 120 umoles/L (2 mg/dl).
**PH designates phenylalanine hydrozylase.
†DHPR designates dihydropteridine reductase.

The low protein natural foods are generally those that are part of a vegetarian diet and include all fruits and most vegetables. Excluded are meats, fish, cheese and other milk products, bread and cake. A system of equivalents has been developed that allows for a relatively simple determination of the amounts of the various permitted foods. Each equivalent contains 15 mg of phe.[12]

Low phe or phe-free preparations. There are two categories of formula used to supplement the diet for PKU.[13] One consists of low phe preparations which are derived from the hydrolysis of proteins such as the casein of milk or albumin. The other category includes preparations which are mixtures of amino acids from which phe is excluded. The low phe formulas are generally used to treat infants and children during the first years of life while the phe-free preparations are used in the treatment of older children and adults.

Nayman *et al*[14] have suggested that the various formulas used in the treatment of the infant with PKU may not be nutritionally optimal since the composition of these preparations is different from that of human milk. They suggested that consideration be given to adjusting these preparations so that they conform to the composition of human milk. This new approach seems to warrant further attention.

Breast feeding. A recent improvement in the treatment of PKU is the successful continuation of breast feeding in infancy. The low phe content of breast milk (41 mg/dl) compared to cow's milk (159mg/dl) has been known for some time.[15] Until recent years, however, the desire for precise assessment of the amount of phe consumed and the general lack of interest in breast feeding among mothers dictated bottle feeding for the management of the infant with PKU. The change towards breast feeding in PKU was inevitable, though, with the increased awareness and education about breast feeding in general and the confidence that resulted from years of monitoring phe levels.

In the past five years several centers have successfully managed breast feeding among infants with PKU. There are two main approaches. In London the infant and mother are admitted to the hospital until control of the blood phe is established. A low phe formula is given before each of five breast feeds following which the infant is allowed to suckle freely. The volume of formula is adjusted according to ten blood phe levels.[16]

In Denver[17] a more rigorous approach has been established. The baby is weighed before and after each breast feeding and the amount of breast milk ingested is determined by the weight gain. The phe requirement for each baby is calculated initially according to published data.[18] Thereafter the intake of breast milk and of the low phe formula are determined by blood phe levels.

III. Dietary Monitoring

Those infants and children with phenylalanine hydroxylase deficiency who are treated with diet must be carefully monitored. This monitoring includes periodic determination of the blood phe level and adjustment of the phe intake according to this level. The acceptable range of blood phe concentration for dietary treatment is 1–10 mg/dl (120–600 umol/l).[19] Most clinics perform specific monitoring weekly or at least semi-monthly. A finger-prick blood specimen obtained by the parent is tested by either the Guthrie assay or the fluorometric method.[20] In addition to biochemical monitoring most centers evaluate the progress of the child with PKU by conducting a periodic assessment of intellectual function and physical development.

IV. Results of Treatment

There is no doubt that control of phenylalanine and related metabolite accumulations by a low phe diet prevents mental retardation in PKU when begun in early infancy.[21,22] The well functioning child or adolescent with PKU who was detected and treated after the advent of routine newborn screening is evidence of this.

On closer examination, however, there often are cognitive difficulties even in the treated child with PKU. The National Collaborative Study of Children Treated for PKU in the United States reported in 1977 that at age four years those treated from the newborn period had I.Q. scores ranging from 12-17 points below those of their parents.[23] More recently Smith *et al* (Smith I. Personal communication, 1981.) have found that children with PKU born within the past few years generally have higher I.Q. scores than similar children who were born a decade ago. A further encouraging note is the report of 1981 from the Collaborative Study in the United States which shows that the mean I.Q. of early treated children with PKU has risen from 93 at age four years to 98 at age six years as determined by testing on the Stanford-Binet Intelligence Scale.[25] Nevertheless, the mean I.Q. of this group is still significantly lower than the mean I.Q. of their non-PKU siblings.

Cognitive difficulties have also manifested themselves in learning disabilities, presumably a result of perceptual-motor dysfunction.[26] The children are often as much as six months behind their chronological age in visual and motor skills.[27] They may be described as hyperactive, confused and distracted,[28] and the results of formal tests suggest that the attention span deficit may be related to a deficit in short-term memory.[29] Difficulties in cognitive problem solving especially in mathematics, are often recognized during early school years.[30]

Emotional and behavioral disturbances are frequently noted, especially among boys.[31] These difficulties extend beyond immature behavior and include those which may be considered antisocial and neurotic. We have encountered suicidal tendencies and even schizoid-type behavior among several adolescent patients with PKU who were treated from early infancy.

V. Discontinuation of Diet

Whether diet can be safely discontinued remains an unanswered question. Some physicians such as Dr. D. Murphy of Dublin (Personal communication, 1979.) and Professor H. Bickel of Heidelberg[33] have advocated life-long adherence to at least some degree of dietary restriction. Others have said that

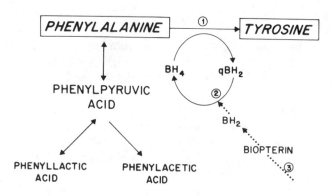

Fig. 1. Pathway of phenylalanine metabolism. Phenylalanine is converted to tyrosine by the activity of phenylalanine hydroxylase.[1] A necessary cofactor for this hydroxylation reaction is tetrahydrobiopterin (BH_4), which is oxidized to quininoid dihydrobiopterin (qBH_2) in the reaction. The availability of sufficient amounts of BH_4 depends on regeneration from qBH_2 by the enzyme dihydropteridine reductase.[2] Since BH_4 is synthesized in the biopterin pathway, a deficiency of biopterin synthesis[3] also results in insufficient BH_4.

Whenever there is a block in the conversion of phenylalanine to tyrosine, whether this results from apoenzyme phenylalanine hydroxylase deficiency as in PKU or from BH_4 deficiency, there is an accumulation of phenylalanine and its metabolites.

---

the diet may be safely discontinued between four and five years of age.[34,35] This policy was followed in a number of centers in the United States.

During the past four years several centers have reported I.Q. scores in children during dietary treatment and following discontinuation of the diet. Smith et al[37] showed that there was a significant drop in I.Q. scores among children in London who had discontinued diet and a downward, though insignificant, trend among Heidelberg children whose diet was relaxed.

When Waisbren et al[38] reviewed the psychological assessments used to determine outcome in the various studies of diet termination and continuation, they noted that there was great variation in the application of the tests, in the types of tests used and in the analysis of data. Unfortunately, any long-term follow-up of PKU is likely to encounter these problems. As these authors pointed out, however, the multicenter data do indicate that some children are seemingly unaffected by diet termination whereas others undergo substantial intellectual deterioration under the same circumstances. Without the means of differentiating these two groups in advance, clinicians are generally recommending continuation of the diet at least through the childhood years.

ISSUES

I. Pterin Defects

An occasional infant with hyperphenylalaninemia, estimated to be from 1–5% of infants detected by newborn screening,[39] will have a deficiency of tetrahydrobiopterin (BH_4). This deficiency can result from reduced activity of dihydropteridine reductase or deficient synthesis of biopterin (Figure 1). We now know that BH_4 is an essential cofactor of phenylalanine hydroxylase, so that when this cofactor is deficient the enzyme will have virtually no activity. BH_4 is also the cofactor of tyrosine-3-hydroxylase and tryptophan-5-hydroxylase, enzymes which lead to the formation of at least three critical neurotransmitters. Thus, the major problem in the pterin defects is lack of neurotransmitter synthesis rather than the elevation of phe, the marker which draws our attention to them.

The pterin defects came to attention as a result of investigations and observations in several centers. From 1958 to 1972 Kaufman et al[40] identified BH_4 as an essential component in the hydroxylation of the aromatic amino acids. In 1975 Smith et al [41] described three children who developed progressive neurological deterioration and died despite early diagnosis and treatment for PKU. They speculated that this condition was probably due to a defect in the metabolism of biopterin in the brain which in turn caused defective synthesis of the neurotransmitters dopamine, norepinephrine and serotonin. Kaufman's group[42] soon confirmed this hypothesis by demonstrating markedly reduced activity of dihydropteridine reductase and decreased levels of neurotransmitter metabolites in a patient with clinical findings similar to those in Smith's original patients. In 1978 Kaufman et al[43] described another patient with hyperphenylalaninemia who had this clinical phenotype. This patient had a defect in the synthesis of biopterin, the site of which is still undefined.

Patients with these defects do not respond to a low phe diet. Therefore, early identification and differentiation from PKU is important if appropriate therapy is to have any chance of success. Several methods for this identification have been used. Liver biopsy with measurement of phenylalanine hydroxylase, dihydropteridine reductase and biopterin confirmed the early cases and is still used for this purpose in some centers.[44] However, less invasive procedures are more applicable for routine use in the investigation of hyperphenylalaninemia.[45] Of these the most promising appears to be measurement of the neopterin: biopterin ratio in urine.[46] A defect in biopterin synthesis results in virtually no urinary biopterin and, therefore, an extremely high ratio. However, a defect in dihydropteridine reductase results in elevated biopterin, hence a low ratio. Dr. E. Naylor of Dr. Guthrie's laboratory in Buffalo is currently providing a national service in the United States for this determination.

It has been proposed that the administration of BH_4 to an individual with hyperphenylalaninemia will differentiate the pterin variants from phe hydroxylase deficiency and would even discriminate between the two pterin defects.[47] In this test the infant is given an oral load of BH_4 and the blood phe is measured. In biopterin synthetase deficiency the blood phe level drops by greater than 50% in eight hours following the load. In hydroxylase deficiency there is little or no change in the blood phe level. It is unclear if in dihydropteridine reductase (DHPR) deficiency the response to BH_4 is similar to that in biopterin synthetase deficiency. Curtius *et al*[47] reported a good response to BH_4 in one patient with DHPR deficiency but this has not been our experience in another patient with this defect (Kaufman S, Levy HL, unpublished data). BH_4 loading of hyperphenylalaninemic neonates is accepted practice in European Centers where the cofactor has been readily available. This cofactor is now authorized for importation from Switzerland (Laboratory of Dr. B. Schircks, CH-8623, Wetzikon, Switzerland) and for use in the United States.

The treatment for the pterin defects has been to give the immediate neurotransmitter precursors: L-dopa and L-5-hydroxtryptophan. This therapy seems to be ineffective in preventing the progressive neurological deterioration of these patients, though it may provide symptomatic benefit. BH_4 replacement therapy is theoretically a more rational form of treatment. However, BH_4 access to the central nervous system is somewhat limited. Nevertheless therapy with large doses of BH_4 or an active analogue which more readily crosses the blood brain barrier might have a future role in the treatment of these defects.[48]

II. Maternal PKU

This is a major problem in PKU. The syndrome of maternal PKU includes mental retardation, microcephaly, congenital heart disease and low birth weight for gestation in offspring of women with PKU. Usually these infants do not have PKU, but they suffer the consequences of intrauterine exposure to the mother's biochemical abnormalities.

The maternal PKU syndrome was recognized as far back as 1957 when Dent reported three nonphenylketonuric but mentally retarded offspring of a mother with PKU.[49] By the mid-1960s Mabry and his colleagues[50] reported additional families and emphasized this as a substantial problem. The number of mentally retarded individuals from this cause could replace those with PKU who would have been retarded were it not for newborn screening and early treatment.[51] This tragedy will be averted only if there is preventive therapy that will allow these women to bear normal children, or, failing this, limitation of their childbearing.

There are many questions about maternal PKU that must be answered:

1. What level of blood phe or degree of biochemical defect in the mother results in fetal damage?
2. Which component(s) of the biochemical phenotype is responsible for the damage? Is it the excess phenylalanine, the phenylalanine metabolites, the relatively low tyrosine, or some other factor?
3. At what stage in pregnancy does the damage occur?
4. Will dietary restriction prevent damage and, if so, when should this restriction begin?
5. How best can girls at risk be counselled and prepared for pregnancy?

Data from a recent international survey[52] indicated that women with classical PKU (blood phe 20 mg/dl or greater) have a 95% risk of having children with microcephaly, a 17% risk of having a child with congenital heart disease, and a 56% risk of having a low birth weight child. When the maternal blood phe level was below 20 mg/dl the risks for mental retardation and microcephaly in offspring were lower. Even when the mother had only mild hyperphenylalaninemia (3–10 mg/dl) there was still a greater than expected frequency of abnormalities in offspring. Ascertainment bias could account for the increased frequency of abnormalities in the latter group since the available data in this survey were insufficient for a definitive measurement of risk. Thus, while it is clear that classical PKU in the pregnant woman constitutes a serious threat to fetal development, we still do not know whether less severe degrees of PKU or hyperphenylalaninemia are equally threatening.

When one is reasonably certain that there is increased risk of fetal damage, the major issue becomes prevention. Perry et al have suggested that these women should be discouraged from becoming pregnant.[53] Aside from the ethical consideration introduced by this suggestion, the reality is that these women desire children and will certainly become pregnant. Consequently, dietary treatment may be the major avenue for preventing fetal damage. There have been fifty or more such treated pregnancies to date.[54,55] The diet has resulted in apparent benefit in a few cases when begun after conception, but in many other cases the offspring have nevertheless been damaged. For instance, in a family that we are following the offspring from pregnancies in which treatment was started during the second trimester are microcephalic and are as reduced in intelligence as their siblings from untreated pregnancies.[56] Smith et al reported severe congenital heart disease in an infant from a pregnancy during which the low phe diet began at seven weeks gestation.[57] By contrast, virtually all reported maternal PKU pregnancies in which diet began prior to conception have resulted in offspring with normal growth and development, at least through the first year of life.[58,59] These reports suggest that to be effective the diet must begin before conception. If so, it is imperative that reproductive counselling be given to girls with PKU so that pregnancies are carefully planned.

At an International Workshop on Maternal PKU in Germany in 1980 the following preliminary recommendations were made:

1. Young girls should not completely discontinue the low phe diet, if possible.
2. Young women with PKU and hyperphenylalaninemia have to be aware of the fact that a pregnancy involves considerable risks.
3. Women who still want to have a child should be counselled and treated dietetically in time, *i.e.* some time before conception.

III. Treatment of the Already Retarded Individual with PKU

One issue that frequently confronts physicians is whether dietary treatment is beneficial to the child who is discovered to have PKU as a result of investigation for mental retardation, or the adult with PKU who is in an institution for the mentally retarded. This question relates to whether the eventual intellectual performance of the child can be improved by even this late treatment, or whether the behavior of the institutionalized adult can be improved and thus become more manageable within the institution.

In the original studies on the treatment of PKU Bickel and his coworkers[60] noted that there were marked improvements in behavior and concentration among mentally retarded young children with PKU who were treated late. Subsequently, others found that the diet produced substantial improvement in the neurological and developmental status of already brain damaged children with PKU if begun before three years of age.[61,62] Conversely, little intellectual benefit seems to result if the diet is begun after the age of three years.[63] However, individual exceptional cases have been reported such as the boy studied by Holmgren *et al*[64] who presented at the age of eight years with an I.Q. of 59, and after one year on the diet had an I.Q. of 82, and who continued to progress while on the diet.

The benefit of dietary treatment to the adolescent or adult with PKU who already has long established brain damage is much more difficult to assess. Behavioral amelioration as a result of dietary therapy has occurred in this group.[65,66] However, a study of this issue using dietary therapy and behavioral modification techniques in institutionalized adults concluded that only the occasional individual is improved.[67] Perhaps it is worthwhile performing a therapeutic trial in any retarded individual with PKU since even slight improvement in a difficult management problem may be worthwhile.

IV. Aspartame

Aspartame is the common name for aspartylphenylalanine methyl ester, a dipeptide consisting of phenylalanine and aspartic acid. This compound is approximately 160 times sweeter than sucrose[68] and is currently in use as an artificial sweetener in several European countries as well as in Canada.

Recently, the U.S. Food and Drug Administration approved this substance for marketing in the United States as a substitute for saccharin.

There has been much controversy in the United States over the safety of this substance for human consumption. A major part of this controversy is the contention that the phenylalanine released in the body following the ingestion of this dipeptide will produce an elevated level of phenylalanine in the body sufficient to be toxic to the brain. Aspartame ingestion does substantially raise the blood phe level in phenylketonuric subjects and it is to be avoided by those who are on a low phe diet.[69] Consequently, labels of food containing aspartame must include a warning to this effect.

Numerous loading studies with aspartame in nonphenylketonuric adults, including those presumed to be heterozygous for the PKU gene, indicate that with even excessively large amounts (100mg/kg) of ingested aspartame the blood phenylalanine level does not rise above a mean concentration of 417 μmoles/liter (6.9 mg/dl).[70,71] In lactating women loading doses of aspartame (50 mg/kg) produced a four-fold rise in blood phenylalanine but the mean peak level remained below 200 μmoles/liter (3.3mg/dl). Breast milk levels of free phenylalanine in these women rose approximately five-fold.[72] It seems reasonable to conclude that aspartame does not represent a threat of brain damage from high phenylalanine levels to the nonphenylketonuric individual or perhaps even to the nonphenylketonuric breast fed infant.

The major remaining unanswered question is whether fetal toxicity might result from aspartame ingestion by pregnant women. From the studies cited above one might conclude that this will probably be no threat when the pregnant woman does not carry the gene for PKU. Also, the woman heterozygous for PKU would presumably not develop a blood phe level that would endanger the nonphenylketonuric fetus unless she ingested vast amounts of aspartame. However, if the fetus is phenylketonuric, or if the pregnant woman has hyperphenylalaninemia, there might be a sufficient accumulation of phenylalanine to result in fetal toxicity. This very important question should be answered as soon as possible.

THE FUTURE

Genetic techniques using the principle of recombinant DNA hold promise for revolutionizing the treatment of inborn errors of metabolism. This may come about in at least two potential ways:

1. The use of these techniques to produce an enzyme such as phenylalanine hydroxylase which could be administered to individuals with PKU.
2. Gene therapy. A normal gene that codes for phenylalanine hydroxylase is inserted into the genome of the phenylketonuric subject and results in endogenous synthesis of this enzyme.[73]

Presently none of this is available, and it may be many years before we can determine whether it is possible or feasible in human therapy. Phenylalanine hydroxylase could become available in large amounts but we may be unable to get it to the site necessary for *in vivo* activity. Alternatively, the enzyme could reach the proper site but be so altered as to lack activity. The introduction of a normal gene which codes for phenylalanine hydroxylase into the phenylketonuric genome could become a reality but the enzyme may not be synthesized due to failure of normal genetic interaction.

Many hurdles must be cleared before the ultimate correction of PKU is attainable. It is less than 50 years since PKU was discovered and the progress in that time has been immense. Perhaps in a future, shorter period of time we will make strides that are unimaginable to us today.

ACKNOWLEDGMENTS

We wish to acknowledge the support of the Sir William Butlin Fellowship from Ireland. This work was also supported by a grant from the National Institutes of Health (NS 05096) and a project grant from the Health Services and Mental Health Administration of the U.S. Public Health Service (01-H-000111).

REFERENCES

1. Garrod AE. Inborn errors of metabolism. The Croonian Lectures. *Lancet* 1908; 2: 1-7, 73-9, 142-48, 214-20.
2. Fölling A. Über Ausscheidung von Phenylbrenztraubensaure in den Harn als Stoffwechselanomalie in Verbindung mit Imbezillitat. *Hoppe Seyler Z Physiol Chem* 1934; 227: 169-76.
3. Bickel H, Gerrard J, Hickmans EM. Influence of phenylalanine intake on phenylketonuria. *Lancet* 1953; 2: 812-13.
4. Guthrie R. Blood screening for phenylketonuria. *J Am Med Assoc* 1961; 178-863.
5. Jervis GA. The genetics of phenylpyruvic oligophrenia. *J Ment Sci* 1939; 85: 719-63.
6. Jervis GA, Block RJ, Bolling D, Kanze E. Chemical and metabolic studies on phenylalanine. II The phenylalanine content of the blood and spinal fluid in phenylpyruvic oligophrenia. *J Biol Chem* 1940; 134: 105-13.
7. Jervis GA. Phenylpyruvic oligophrenia. Deficiency of phenylalanine-oxidizing system. *Proc Soc Exp Biol Med* 1953; 82: 514-15.
8. MacCready RA. Phenylketonuria in the newborn. *Lancet* 1963; 2: 46.
9. Deleted in editing.
10. Deleted in editing.
11. Levy HL. Genetic screening *In:* Harris HH, Hirschhorn K, eds. Advances in human genetics. Vol 4. New York: Plenum Press, 1973: 1-104.
12. Acosta PB, Wenz E, Schaeffler G, Koch R. PKU-A Diet Guide for parents of children with phenylketonuria. Evansville: Mead Johnson Laboratories. 1969.
13. Pueschel SM, Hum C, Andrews M. Nutritional management of the female

with phenylketonuria during pregnancy. *Am J Clin Nutr* 1977;30: 1153-61.
14. Nayman R, Thomson ME, Scruier CR, Clow CL. Observations on the composition of milk substitute products for treatment of inborn errors of amino acid metabolism. Comparisons with human milk. A proposal to rationalize nutrient content of treatment products. *Am J Clin Nutr* 1979; 32: 1279-89.
15. Macy IG. Composition of human colostrum and milk. *Am J Dis Child* 1949; 78: 589-603.
16. Deleted in editing.
17. Earnest AE, McCabe ERB, Neifert MR, O'Flynn ME. Guide to breast feeding the infant with PKU. DHHS publication no. (HSA) 79-5110. Washington: U.S. Government printing office, 1980.
18. Acosta PB, Wenze E. Diet management of PKU for infants and preschool children. DHHS publication no. (HSA) 77-5209. Washington: U.S. Government printing office, 1977.
19. Williamson ML, Koch R, Azen BH, Chang C. Correlates of intelligence tests results in treated phenylketonuric children. *Pediatrics* 1981; 68: 161-67.
20. McCaman MW, Robins E. Fluorometric method for the determination of phenylalanine in serum. *J Lab Clin Med* 1962; 59: 885-90.
21. Kang ES, Sollee ND, Gerald PS. Results of treatment and termination of the diet in phenylktonuria (PKU). *Pediatrics* 1970; 46: 881-90.
22. Smith I, Wolff OH. Natural history of phenylketonuria and influence of early treatment. *Lancet* 1974; 2: 540-44.
23. Dobson JC, Williamson ML, Azen C, *et al.* Intellectual assessment of 111 four-year-old children with phenylketonuria. *Pediatrics* 1977; 60: 822-27.
24. Deleted in editing.
25. Williamson ML, Koch R, Azen C, Chang C. Correlates of intelligence test results in treated phenylketonuric children. *Pediatrics* 1981; 68: 161-67.
26. Koff E, Boyle P, Pueschel SM. Perceptual-motor functioning in children with phenylketonuria. *Am J Dis Child* 1977; 131: 1084-87.
27. Ohen A, Laubi T, Gitzelmann R. Perceptual motor testing in children with phenylketonuria at age 13 months to 12 years. *Eur J Pediatr* 1979; 130: 208.
28. Sutherland BS, Umbarger B, Berry HK. The treatment of phenylketonuria; A decade of results. *Am J Dis Child* 1966; 3: 505-23.
29. Francois B, Willems G. Use of pemolin in the treatment of learning disorders in children with phenylketonuria (PKU). *Pediatr Res* 1980; 14: 174.
30. Berry HK, O'Grady DJ, Perlmutter LJ, Bofinger MK. Intellectual development and acadamic achievement of children treated early for phenylketonuria. *Devel Med Child Neurol* 1979; 21: 311-20.
31. Stevenson JE, Hawcroft J, Lobascher M, *et al.* Behaviorial variance in children with early treated phenylketonuria. *Arch Dis Child 1979; 54: 14-8.*
32. Deleted in editing.
33. Bickel H. Phenylketonuria: Past, present, future. *J Inher Metab Dis* 1980; 3: 123-32.
34. Holtzman NA, Welcher DW, Mellets ED. Termination of restricted diet in children with phenylketonuria: A randomised controlled study. *N Engl J Med* 1975; 293- 1121-24.
35. Kogf E, Kammerer B, Boyle P, Pueschel SM. Intelligence and phenylketonuria: effects of diet termination. *J Pediatr* 1979; 94: 534-7.
36. Cabalska B, Duczynska N, Borzymowska J, *et al.* Termination of dietary treatment in phenylketonuria. *Eur J Pediatr* 1977; 126: 253-62.

37. Smith I, Lobascher ME, Stevenson JE, et al. Effect of stopping low-phenylalanine diet on intellectual progress of children with phenylketonuria. Brit Med J 1978; 2: 723-26.
38. Waisbren SE, Schnell RR, Levy HL. Diet termination in children with phenylketonuria: A review of psychological assessments used to determine outcome. J Inher Metab Dis 1980; 3: 149-53.
39. Berlow S. Progress in phenylketonuria: Defects in the metabolism of biopterin. Pediatrics 1980; 65: 837-39.
40. Kaufman S. Phenylketonuria: Biochemical mechanisms. In: Agranoff BW, Aprison MH, eds. Advances in neurochemistry. Vol. 2. New York: Plenum Press, 1976: 1-132.
41. Smith I, Clayton BE, Wolff OH. New variant of phenylketonuria with progressive neurological illness unresponsive to phenylalanine restriction. Lancet 1975; 1: 1108-11.
42. Kaufman S, Holtzman A, Milstien S, et al. Phenylketonuria due to a deficiency of dihydropteridine reductase. N Engl J Med 1975; 293: 785-90.
43. Kaufman S, Berlow S, Summer GK, et al. Hyperphenylalaninemia due to deficiency of biopterin. N Engl J Med 1978; 299: 673-9.
44. Berry HK. The diagnosis of phenylketonuria. Am J Dis Child 1981; 135: 211-13.
45. Kaufman S. Differential diagnosis of variant forms of hyperphenylalaninemia. Pediatrics 1980; 65: 840-42.
46. Nixon JC, Lee C−L, Milstien S, et al. Neopterin and biopterin levels in patients with atypical forms of phenylketonuria. J Neurochem 1980; 35: 898-904.
47. Curtius H-CH, Niederwieser A, Viscontini M, et al. Atypical phenylketonuria due to tetrahydrobiopterin deficiency: Diagnosis and treatment with tetrahydrobiopterin, dihydrobiopterin and sepiapterin. Clin Chim Acta 1979; 93: 251-62.
48. Kapatos G, Kaufman S. Peripherally administered reduced pterins do enter the brain. Science 1981; 212: 955-56.
49. Dent CE. Discussion of Armstrong MD: Relation of biochemical abnormality to development of mental defect in phenylketonuria. In: Etiologic factors in mental retardation: Report of Twenty-third Ross Paediatric Research Conference, November 8-9, 1956. Columbus, Ohio: Ross Laboratories, 1957: 32-3.
50. Mabry CC, Denniston JC, Nelson TL, et al. Maternal phenylketonuria: A cause of mental retardation in children without the metabolic defect. N Engl J Med 1963; 269: 1404-08.
51. Kirkman HN, Jr. Projections of mental retardation from PKU. Pediatr Res 1979; 13: 414.
52. Lenke RR, Levy HL. Maternal phenylketonuria and hyperphenylalaninemia. An international survey of the outcome of untreated and treated pregnancies. N Engl J Med 1980; 303: 1202-08.
53. Perry TL, Hansen S, Tischler B, et al. Unrecognized adult phenylketonuria. Implications for obstetrics and psychiatry. N Engl J Med 1973; 289: 395-98.
54. Lenke RR, Levy HL. Maternal phenylketonuria. Results of dietary therapy. Am J Obstet Gynecol, in press.
55. Bickel H, ed. Maternal phenylketonuria. Frankfurt: Maizena, 1980.
56. Levy HL, Kaplan GN, Erickson AM. Comparison of untreated and treated pregnancies in a mother with phenylketonuria. J Pediatr, in press.
57. Smith I, Macarthey FJ, Endohazi M, et al. Fetal damage despite low-

phenylalanine diet after conception in a phenylketonuric woman. *Lancet* 1979; 1: 17-9.

58. Nielsen, KB, Guttler F, Wever J, *et al.* Dietary treatment of a phenyl-ketonuric woman during planned pregnancy. *In*: Bickel H, ed. Maternal phenylketonuria. Frankfurt: Maizena. 1980; 87-91.

59. Brenton DB, Cusworth DC, Garrod P, *et al.* Maternal phenylketonuria treated by diet before conception. *In*: See Ref. 58. pp. 67-71.

60. Bickel H, Gerrard J, Hickmans EM. The influence of phenylalanine intake on the chemistry and behavior of a phenylketonuric child. *Acta Paediat Scand* 1954; 43: 64-77.

61. Centerwall WR, Centerwall SA, Armon V, *et al.* Phenylketonuria. II. Results of treatment of infants and young children. *J Pediatr* 1961; 59: 102-18.

62. Knox WE. Phenylketonuria. *In*: Stanbury JB, Wyngaarden JB, Fredrickson DS, eds. The metabolic basis of inherited disease. Second edition. New York: McGraw-Hill, 1966: 258-94.

63. Hsia D Y-Y, Know WE, Quinn KV, *et al.* A one year control study of the effect of low-phenylalanine diet on phenylketonuria. *Pediatrics* 1958; 21: 178-202.

64. Holmgren G, Blomquist HK, Samuelson G. Positive effect of a late introduced modified diet in an 8-year-old PKU child. *Neuropediat* 1980; 11: 76-9.

65. Hambaeus L, Holmgren G, Samuelson G. Dietary treatment of adult patients with phenylketonuria. *Nutr Metab* 1979; 13: 298-317.

66. Anderson VE, Siegel FS, Bruhl HH. Behavioral and biochemical correlates of diet change in phenylketonuria. *Pediatr Res* 1976; 10: 10-7.

67. Marholin D, Pohl RE, Stewart RM, *et al.* Effects of diet and behavior therapy on social and motor behavior of retarded phenylketonuric adults. An experimental analysis. *Pediatr Res* 1978; 12: 179-87.

68. Clowinger MR, Baldwin RE. Aspartylphenylalanine methyl ester: A low-calorie sweetener. *Science* 1970; 179: 81-2.

69. Koch R, Schaeffler G, Shaw KNF. Results of loading doses of aspartame by two phenylketonuric (PKU) children compared with two normal children. *J Toxicol Environ Hlth* 1976; 2: 459-69.

70. Stegink LD, Filer LJ Jr, Baker GL, *et al.* Effect of aspartame loading upon plasma and erythrocyte levels in phenylketonuric heterozygotes and normal adult subjects. *J Nutr* 1979; 109: 708-17.

71. Stegink LD, Filer LJ Jr, Baker GL, *et al.* Effect of an abuse dose of aspartame upon plasma and erythrocyte levels of amino acids in phenylketonuric heterozygous and normal adults. *J Nutr* 1980; 110: 2216-24.

72. Stegink LD, Filer LJ Jr, Baker GL. Plasma, erythrocyte and human milk levels of free amino acids in lactating women administered aspartame or lactose. *J Nutr* 1979; 109: 2173-81.

73. Miller WL. Recombinant DNA and the pediatrician. *J Pediatr* 1981; 99: 1-15.

ENZYMOLOGICAL DIAGNOSIS OF LYSOSOMAL STORAGE DISORDERS

J.A. Lowden
M.A. Skomorowski
P.M. Strasberg
A. Novak

It is appropriate in 1981 to include a session on lysosomal storage disease (LSD) in a symposium on clinical genetics because it is now 100 years since Warren Tay described the cherry red spot in the fundus of a retarded child, thereby lending his name to the first description of a lysosomal storage disease, *i.e.* Tay-Sachs Disease.[1] In the intervening years a host of clinical descriptions followed and today the lysosomal disorders can be classified into three broad categories resulting from defects in the activity of more than two dozen enzymes (Table 1). Many of these disorders occur in a variety of clinical subtypes and some, like the GM_2 gangliosidoses, are the result of defects in more than one gene.[43]

The simplest classification divides the LSDs into three subgroups: the sphingolipidoses, the mucopolysaccharidoses and the oligosaccharidoses. Table 1 indicates the year in which each disorder was first described in the literature. While most of the original clinical descriptions were published over 50 years ago, the current literature abounds with new reports of clinical variants of all of these diseases. New enzyme defects also continue to be added to the list. Salla disease was described as recently as 1979.[42]

Although the clinical reports date back over the last century, and the first descriptions of the nature of the storage compounds appeared more than forty years ago, it was in the mid 1960s that techniques for demonstrating the enzymopathies became available. These techniques have had a profound effect on the field. For example, the demonstration of the defect in hexosaminidase A activity in Tay-Sachs disease in 1969[2] led to the establishment of carrier screening programs by 1970[44] and virtual elimination of the disease in Jewish families today.[45] As a corollary to this change there has been an incredible expansion in the numbers of recognized clinical phenotypes resulting from hexosaminidase defects. We now diagnose not only infantile but juvenile and adult forms of the disease[46] and have come to expect movement disorders to appear as a major sign in older patients.[47,48] Animal models of hexo-

TABLE 1
Lysosomal Storage Diseases — First Descriptions

Class	Name	Clinical Description	Enzymatic Defect*
Sphingolipidoses	GM_2 Gangliosidosis	1881[1]	1969[2]†
	GM_1 Gangliosidosis	1965[3]	1968[4]
	Metachromatic Leucodystrophy	1910[5]	1963[6]
	Globoid cell Leucodystrophy	1916[7]	1970[8]
	Gaucher	1882[9]	1965[10]
	Niemann-Pick	1914[11]	1966[12]†
	Farber's Lipogranulomatosis	1957[13]	1975[14]
	Fabry's	1898[15]	1967[16]
	Multiple Sulfatase Def	1973[17]	1973[17]
Mucopoly-	I Hurler	1920[18]	1972[19]
saccharidoses	Scheie	1962[20]	1972[19]
	II Hunter	1917[21]	1973[22]
	III Sanfilippo	1963[23]	1972[24]†
	IV Morquio	1935[25]	1974[26]†
	VI Maroteaux-Lamy	1963[27]	1974[28]
	VII β-Glucuronidase Def	1973[29]	1973[29]
Oligo-	Sialidosis	1968[30]	1977[31]†
saccharidoses	I-Cell Disease	1967[32]	1981[33]
	Mucolipidosis IV	1974[34]	1979[35]
	Mannosidosis	1967[36]	1970[37]
	Fucosidosis	1966[38]	1968[39]
	Aspartylglucosaminuria	1967[40]	1968[41]
	Salla Disease	1979[42]	—

*In some cases the enzyme defect was described simultaneously in more than one laboratory. For simplicity only a single reference is given.

†These disorders were subsequently demonstrated to have several sub-classifications with more than one gene defect.

saminidase defects have aided in our understanding of pathophysiology[49] and the field is far more complex than Tay or Sachs ever imagined. This one example illustrates how a single disease can appear with many phenotypes, many gene disorders and many storage compounds. Many other lysosomal disorders are similar.

Expansion of the diagnostic horizons for each of several enzymopathies has brought many problems to the diagnostic laboratory and it is the nature of those problems that will be discussed in this paper. While probably 90–95% of patients with LSDs can be identified with a high degree of suspicion on clinical grounds, can be subclassified with a few simple tests, and can be readily diagnosed by measuring the appropriate lysosomal hydrolase,[11] a small core of patients will be missed or misdiagnosed because of the vagaries of these particular enzymes. What are the more important pitfalls in the brand of enzymology and where will the decisive answers be found?

THE NATURE OF LYSOSOMAL HYDROLASES

In 1949, de Duve and his colleagues[51] uncovered a particle-entrapped form of acid phosphatase which could be released upon *in vitro* storage. Their observations led to the recognition that the lysosome was a sub-cellular organelle containing many hydrolases. The observed enzymes have now been shown to be glycoproteins containing branched, high mannose or complex type oligosaccharide side chains. Defects in the processing of this carbohydrate part of the enzymes prevent proper binding of the hydrolase within the lysosome and result in I-cell disease.[33,52] In this disorder many lysosomal enzymes are decreased in activity due to a single defect in the gene responsible for such processing. However, loss of activity of a single hydrolase is seen in most lysosomal storage disorders and leads to storage of the macromolecular substrate(s) of that enzyme in the lysosome. The stored material may have a characteristic structure like the whorled membranes seen in ganglioside storage,[53] or may be amorphous like that found in sialidosis.[54] A defect in the activity of one hydrolase is often associated with striking increases in the activity of other hydrolases.[55] Most lysosomal enzymes have m.w. of 100,000–200,000 daltons, but they primarily occur in multiple forms as various-sized multiples of a smaller polypeptide or polypeptides. Many of the hydrolases require an associated non-catalytic activator or co-hydrolase for cellular activity. These co-molecules are small (10–20,000 daltons) polypeptides and will be discussed in more detail later.

Difficulties in accurate assessment of the role of a specific lysosomal hydrolase in the pathogenesis of a storage disease often result from problems associated with *in vitro* assays of enzymes. They include:

1. Differences between natural versus synthetic substrates.
2. Requirements for natural activator or co-hydrolase proteins.

TABLE 2

Conditions of Hydrolysis (Synthetic Substrates) for Lysosomal Hydrolases in Cultured Fibroblasts

Enzyme	Substrate	Conc. (mM)	Buffer (M)	pH	Protein (μg/assay)	Final volume (μl)
β-gal.	4MU-β-Gal	0.5	0.04 CP*	4.1	10-25	1050
β-hex.	4MU-β-GlcNAc	5.0	0.04 CP	4.1	1-5	200
β-gluc.	4MU-β-Glc	7.0	0.2 NaAc	5.5	50-100	200
α-neur.	4MU-α-NeuNAc	4.0	0.5 NaAc	4.3	30-50	30
α-gluc.	pNP-α-GlcNAc	4.0	0.4 CP	4.5	200-1500	200
α-mann.	4MU-α-Man	1.0	0.4 CP	4.4	50-150	400
α-fuc.	4MU-α-Fuc	1.0	0.4 CP	5.5	5-25	110
α-idur.	phenyliduronide	10	0.4 Na formate	3.5	50-100	75
arylsulf. A	4-nitrocatechol SO$_4$	10	0.5 NaAc	5.0	10-50	300
arylsulf. B	4-nitrocatechol SO$_4$	62	0.5 NaAc + 0.1 BaAc	6.0	75-150	250

*Citrate-phospate buffer according to McIvaine.

3. Presence of lysosomal and non-lysosomal hydrolases.
4. Secondary effects on other enzymes resulting from the primary gene defect.
5. The vagaries of demonstrating a gene-dose effect in heterozygotes.

Many other problems can trouble the enzymologist who wishes to establish a diagnosis for a lysosomal disorder, but the above list serves to make the point. Be careful when assaying the activity of these enzymes. Use proper controls and be certain you know what you are doing.

NATURAL VERSUS SYNTHETIC SUBSTRATES

Lysosomal hydrolases may be analyzed using either the natural substrate for the particular enzyme or one of a number of synthetic substrates. The commonest of the latter are the p-nitrophenyl or 4-methylumelliferyl (4MU) derivatives. As an example, arylsulfatase activity can be measured with 4-nitrocatechol-sulfate (NCS). The conditions for assay vary from one enzyme to another, and those used in this laboratory are listed in Table 2. We employ 4MU substrates almost exclusively because of their high sensitivity and the resultant ease of such assays. Although many laboratories use various alkaline glycine buffers to stop the reaction and increase the measurable end fluorescence, we have routinely used amino-methyl propanol (0.1M, pH 10.4) because of its greater stability. This reagent is prepared in 10 fold excess concentration and stored on the bench for many months without change. Fluorescence is read in a Turner spectrofluorometer (excitation 365 nm, emission 450 nm).

Conditions for natural substrate analyses also vary from one laboratory to another. The substrates are largely unavailable commercially and must be isolated and/or radiolabelled in the laboratory. Some, like GM_2 ganglioside, can only be isolated in adequate quantities from pathological tissue samples and thus can be prepared in few laboratories. Therefore, we employ only a select number of natural substrates. Our conditions are listed in Table 3. Natural substrates are usually added in the presence of detergents. The assays often involve the release of a radioactively labelled group from the substrate. The natural substrate techniques are more costly, and more time-consuming; however, they are more specific. Problems with natural substrate assays arise in the interpretation of the results. The added detergents give distorted kinetics due to interaction with the substrate,[63] or have direct effects on the enzyme itself.[64] In GM_2-gangliosidase assays, the normally low enzyme activity encourages the use of prolonged hydrolysis times, further distorting the kinetics. Additionally,[+] because most of the natural substrates have a large non-polar component, they tend to form micelles which may further distort the kinetics.[65] Finally, natural substrates are often difficult to solubilize, and may

+See Table 4.

TABLE 3
Conditions of Hydrolysis (Natural Substrates) for Lysosomal Hydrolases in Cultured Fibroblasts

Enzyme	Substrate	Conc. (mM)	Buffer (mM)	pH	Protein (μg)	Final volume (μl)	Ref.
β-gal.	GM$_1$ ganglioside (^{3}H-Gal)*	0.3	100 NaAc	4.3	10-25	100	56
β-hex.	GM$_2$ ganglioside (^{3}H-GalNAc)*	0.6	15 CPC	4.1	50.100	50	57
β-gluc.	Glucocerebroside (^{3}H-ceramide)†	1.2	75 NaCt	5.6	50-100	200	58
Sulf.	Galactocerebroside SO$_4$ (^{3}H-ceramide)† ●	0.35	12 NaAc	5.0	90-120	200	**
Sphingo.	Sphingomyelin (^{3}H-ceramide)†	0.18	50 NaAc	5.0	20-50	200	59
Galactocere. ●	Galactocerebroside (^{3}H-gal)*	0.40	50 CP$^+$	4.2	100-300	200	56

*Prepared by reduction with Na borotritide after treatment with galactose oxidase.[61]

†Prepared by reduction of sphingosine to dihydrosphingosine in presence of tritium gas.[62]

+CP = citrate-phospate buffer prepared according to McIvaine.

**Skomorowski MA, Lowden JA, Unpublished method.

TABLE 4

Comparison of Enzyme Activity with Natural vs Synthetic Substrate in Cultured Fibroblasts

Enzyme	Natural substrate*	Activity (nmoles/mg protein/hr)	Synthetic substrate†		Ratio syn:nat
β-gal.	^{3}H-GM$_1$ ganglioside	72	4MU-β-galactoside	333	4:6
β-gluc.	^{3}H-glucosylceramide	4.1	4MU-β-glucoside	202	49:8
Sulf.	^{3}H-galactocerebroside-SO$_4$	16.8	4-nitrocatechol SO$_4$	564	33:6
β-hex. A	^{3}H-GM$_2$ ganglioside	6.4	4MU-β-glucosaminide	6100	95:3

*Conditions of hydrolysis given in Table 3.

†Conditions of hydrolysis given in Table 2.

+A comparison of the ratios of synthetic to natural substrates shows that all natural substrates are hydrolysed more slowly than their water-soluble synthetic counterparts, but some like GM$_2$ ganglioside are almost not cleaved.

43

require more detergents; it may be impossible to measure activity at a substrate concentration as high as the Km.

However, there are many good arguments for using natural substrates. Sphingomyelinase defects have been studied with synthetic substrates, but Jones et al[66] in a detailed study determined that neither Bis(−4MU−) phosphate nor hexadecanoyl (nitrophenyl) phosphorylcholine gave similar kinetic parameters to use of natural substrate with purified sphingomyelinase and neither were suitable for the diagnosis of sphingomyelinase deficiency states. Further, some patients with GM_2 storage disease appear to be genetic compounds with one parent who tests as a normal while the other tests like a classical Tay-Sachs carrier, using 4MU substrates.[67] For clarification natural substrate analyses are obviously important. Recently Gatt et al[68] have developed a sphingomyelin analogue in which the fatty acid on the ceramide portion of the molecule has been substituted with fluorescent lipid (anthroyl-oxy-undecanoyl). These artificial substrates can theoretically be developed for any sphingolipid and may prove invaluable in the study of "natural" substrate hydrolysis.

ACTIVATORS OR COHYDROLASES

Studies by Fischer and Jatzkewitz[69] and later by Li et al[70] suggested that many lysosomal enzymes required an associated protein activator or effector substance for hydrolysis of the natural substrate. Usually these activators can be replaced in in vitro assay systems by using detergents. However, it is clear that there are additional important components in the hydrolysis of sulfatide,[69] glucocerebroside[71] and GM_2 ganglioside.[64] Specific genetic defects, in which enzyme activity appears normal with the artificial substrate but is defective when the natural substrate is used, have been demonstrated only for cerebroside sulfatidase[72] and GM_2 gangliosidase.[64] A recent report indicating that Type C Niemann-Pick disease results from an activator defect awaits confirmation.[73] If activators are important for lysosomal enzyme activity we should expect to find genetic defects for each activator polypeptide.

Metachromatic leukodystrophy is usually diagnosed by a defect in the activity of arylsulfatase A activity, using 4-nitrocatechol sulfate (NCS). However, two new reports suggest that the specific substrate, sulfatide, should be used in some cases. Shapiro et al[72] have described a patient who had normal arylsulfatase A activity using the NCS substrate but whose cells could only catabolize the natural substrate when they were incubated with the specific activator.

A similar picture presents with some hexosaminidase mutants. The AB variant patients have normal activity for both heat-stable and heat-labile hexosaminidases using 4MU substrates but cannot catabolize GM_2 ganglioside.[43] Their fibroblasts can hydrolyse the natural substrate in the presence of activator protein.[64]

MIXTURES OF LYSOSOMAL AND NON-LYSOSOMAL HYDROLASES

The literature on lysosomal enzyme defects abounds with data on patients who appear to have 10–25% of normal levels of activity although they clearly have present a clinical disease state and have been shown to store the particular substrate.[74-76] In some instances this difference is due to an activator defect (see above), but in most it probably represents the activity of non-lysosomal enzyme with a different pH optimum. Perhaps the problem is most clearly seen with β-glucosidase. This somewhat confusing enzyme activity has a lysosomal form with a pH optimum of 4.5 and a neutral form with a pH optimum of 5.5–6.0. When assayed in the presence of sodium taurocholate the pH optimum of the lysosomal enzyme shifts to 5.0. The neutral enzyme is inhibited by taurocholate. Thus, enzymic activity in Gaucher tissues may appear as a significant percentage of normal if assayed at an inappropriate pH or in the absence of taurocholate.[77]

Similarly, neutral β-galactosidase can confuse the diagnosis of GM_1 gangliosidosis and may require the use of natural substrates to establish the enzymopathy.[78]

PRIMARY VERSUS SECONDARY GENE DEFECTS

When measuring lysosomal enzyme activity one can frequently make errors in diagnosis by failing to properly assess the patient, or by not demonstrating the heterozygote status of the parents. On one occasion some years ago we received a leucocyte pellet from another center. The physician asked us to confirm a diagnosis of Hurler syndrome. We measured the α-iduronidase and found less than 1% of normal activity. However, we found similar low levels of arylsulfatase and β-galactosidase. Analysis of serum enzymes and later of enzymes in cultured fibroblasts confirmed the diagnosis of I-cell disease. In that disorder most lysosomal enzymes are found to have low activity in cells while serum levels are greatly elevated.

In a similar fashion, the first patient we diagnosed with β-gal$^-$/neur$^-$ sialidosis[34] was thought clinically to have GM_1 gangliosidosis. She resembled a typical patient with the infantile form, and when the leucocyte β-galactosidase was absent we thought we had confirmed the diagnosis. Her parents, however, had normal β-galactosidase activities in leucocytes. Subsequent examination of urinary oligosaccharides eventually led to a diagnosis of sialidosis. Fibroblast α-neuraminidase in the parents' cells showed the expected gene dose effect and confirmed the diagnosis (Table 5). In this patient the primary gene defect was in neuraminidase. β-galactosidase is affected in a secondary manner which is not clearly understood but which may make that enzyme more susceptible to proteolysis.[80]

TABLE 5

Demonstration of Primary Gene Defect in β-Gal⁻/α-Neur⁻ Sialidosis

Patients	Leukocytes β-Gal	Fibroblasts β-Gal	α-Neur
1	0.04	–	–
2	8.4	52	0
3	12.0	70	0
Parents	180±47 (6)	336±83 (5)	73±10 (5)
Controls	186±51	395±64	126±22

Values are means ± SD calc. as mg 4-methylumbilliferone released/mg protein/hr.

Note: While the affected infants all had defects in the activity of both β-galactosidase and α-neuraminidase, the parents only showed a decreased activity for α-neuraminidase.

CARRIER DETECTION

In most single gene disorders, one expects the obligate heterozygotes to have half-maximal enzyme activity for the particular enzyme in question. But the gene dose effect cannot always be demonstrated. In a review of data from several laboratories Shapiro[81] found that while it was reasonably safe to attempt to identify Tay-Sachs heterozygotes, most other lysosomal enzymes cannot yet be assayed as reliable indicators of heterozygote status. There are many reasons for this unreliability including:

1. Variations in or lack of standardization of methods.
2. Variations in enzyme source (*e.g.* serum vs plasma, cell culture conditions, etc.).
3. Effect of non-genetic factors (*e.g.* oral contraceptives and liver disease in Tay-Sachs carrier tests.[82])
4. Effect of other non-structural genes on activity of a particular enzyme.
5. Variant alleles at the gene locus.
6. Extreme lyonization in carriers of X-linked disease.
7. Small sample sizes studied by most investigators.

The success of the Tay-Sachs carrier screening programs[45] indicates that we must persist in our attempts to identify heterozygotes, but as Dr. Kazazian has shown so clearly (see chapter "Prenatal Diagnosis of Hemoglobinopathies by Restriction Analysis: Methodology and Experience"), the future may provide better ways to identify carriers than by measuring enzyme activity.

Currently, the problems of carrier detection are the same problems of measuring lysosomal enzymes for making diagnoses in patients. Also, in carrier detection the relative precision must be greatly improved. Indeed, for some enzymopathies there are overlaps between patients and carriers with low levels of enzyme activity[83] but the overlaps between normals and carriers are far from common.

Because lysosomal enzymopathies are recognized at an ever-increasing rate by more suspicious clinicians, we will continue to measure these enzymes for diagnostic purposes. In time the techniques of molecular biology may override the need to conduct this form of clinical enzymology, but that day is still quite distant. Until it arrives we must continue to use great care in diagnosis. Technologies may improve and our understanding of allelic mutants will expand. However, in the laboratory strict adherence to careful technique, to the use of appropriate controls and to an attitude of constantly suspecting the unusual, will result in provision of the best service for physicians and the patients and families they counsel.

ACKNOWLEDGMENT

This work was supported in part by Program Grant PG-4 from the Medical Research Council of Canada.

REFERENCES

1. Tay W. Symmetrical changes in the region of the yellow spot in each eye of an infant. *Trans Ophthalmol Soc UK* 1881; 1:55-57.
2. Okada S, O'Brien JS. Tay-Sachs disease: Generalized absence of a beta-D-N-acetylhexosaminidase component. *Science* 1969; 165: 698-700.
3. Landing BH, Silverman FN, Craig JM, Jacoby MD, Lahey ME, Chadwick DL. Familial neurovisceral lipidosis: An analysis of eight cases of a syndrome previously reported as "Hurler variant", "pseudo-Hurler disease", and "Tay-Sachs disease with visceral involvement". *Am J Dis Child* 1964; 108: 503-22.
4. Okada S, O'Brien J. Generalized gangliosidosis: Beta-galactosidase deficiency. *Science* 1968; 160: 1002-04.
5. Alzheimer A. Beitrage zur Kenntnis der pathologischen Neuroglia und ihrer Beziehungen zu den Abbauvorgangen im Nervengewebe. *Nissel-Alzheimer's Histol Histopathol Arb* 1910; 3: 493.
6. Austin JH, Balasubramanian AS, Pattabiraman TN, Saraswathi S, Basu DK, Bachawat BK. A controlled study of enzymic activities in three human disorders of glycolipid metabolism. *J Neurochem* 1963; 10: 805-16.
7. Krabbe J. A new familial infantile form of diffuse brain sclerosis. *Brain* 1916; 39: 74.

8. Suzuki K, Suzuki Y. Globoid cell leucodystrophy (Karbbe's disease) deficiency of galactocerebroside β-galactosidase. *Proc Nat Acad Sci USA* 1970; 66: 302-09.
9. Gaucher PCE, De l'epithelioma primitif de la rate. These de Paris. 1882.
10. Brady RO, Kanfer JN, Shapiro D. Metabolism of glucocerebrosides II. Evidence of an enzymatic deficiency in Gaucher's disease. *Biochem Biophys Res Commun* 1965; 18: 221-25.
11. Niemann A. Ein unbekanntes Krankheitsbild. *Jahrb Kinderheilkd* 1914; 29: 1.
12. Brady RO, Kanfer JN, Mock MB, Fredrickson DS. The metabolism of sphingomyelin II. Evidence of an enzymatic deficiency in Niemann-Pick disease. *Proc Nat Acad Sci USA* 1966; 55: 366-69.
13. Farber S, Cohen J, Uzman LI. Lipogranulomatosis: A new lipo-glycoprotein "storage" disease. *J Mt Sinai Hosp* 1957; 24: 816-37.
14. Sugita M, Dulaney JT, Moser HW. Ceramidase deficiency in Farber's disease (lipogranulomatosis). *Science* 1975; 178: 1100-02.
15. Fabry J. Ein Beitrag zur Kenntniss der Purpura haemorrhagica nodularis. *Arch Dermat Syph* 1898; 43: 187-200.
16. Brady RO, Gal AE, Bradley RM, Martensson E, Warchaw L, Laster L. Enzymic defect in Fabry's disease: Ceramide trihexosidase deficiency. *New Eng J Med* 1967; 276: 1163-67.
17. Austin J. Studies in metachromatic leukodystrophy XII: Multiple sulfatase deficiency. *Arch Neurol* 1973; 28: 258.
18. Hurler G. Uber einen Typ multipler. Abartungen, vorwiegend am Skelettsystem. *Ztschr Kinderh* 1920; 24:220.
19. Bach G, Friedman R, Weissman B, Neufeld EF. The defect in the Hurler and Scheie syndromes: Deficiency of α-L-iduronidase. *Proc Nat Acad Sci USA* 1972; 69: 2048-51.
20. Scheie HG, Hambrick Jr. GW, Barness LA. A newly recognized forme fruste of Hurler's disease (gargoylism). *Am J Ophthalmol* 1962; 53: 753-69.
21. Hunter C. A rare disease in two brothers. *Proc R Soc Med* 1917; 10: 104-16.
22. Bach G, Eisenberg Jr. F, Cantz M, Neufeld EF. The defect in the Hunter syndrome: Deficiency of sulfoiduronate sulfatase. *Proc Nat Acad Sci USA* 1973; 70: 2134-38.
23. Sanfilippo S, Podosin R, Langer L, Good RA. Mental retardation associated with acid mucopolysacchariduria (heparitin sulfate type). *J Pediatr* 1963; 63: 837-38.
24. O'Brien JS. Sanfilippo syndrome: Profound deficiency of α-acetylglucosaminidase activity in organs and skin fibroblasts from Type-B patients. *Proc Nat Acad Sci USA* 1972; 69: 1720-22.
25. Morquio L. Sur une forme de dystrophie osseuse familiale. *Arch Med Enf* 1935; 38: 5-24.
26. Matalon R, Arbogast B, Justice P, Brandt I, Dorfman A. Morquio's syndrome: Deficiency of a chondroitin sulfate N-acetylhexosamine sulfate sulfatase. *Biochem Biophys Res Commun* 1974; 61: 709-15.
27. Maroteaux P, Leveque B, Marie J, Lamy M. Une nouvelle dysostose avec elimination urinaire de chondroitine-sulfate B. *La Presse Med* 1963; 71: 1849-52.
28. Fluharty AL, Stevens RL, Sanders DL, Kihara H. Arylsulfatase B deficiency in Maroteau-Lamy syndrome cultured fibroblasts. *Biochem Biophys Res Commun* 1974; 59: 455-61.

29. Sly WS, Quinton BA, McAlister WH, Rimoin DL. β-glucuronidase deficiency: Report of clinical, radiologic and biochemical features of a new mucopolysaccharidosis. *J Pediatr* 1973; 82: 249-57.
30. Spranger J, Wiedemann HR, Tolkdorf M, Graucob E, Caesar R. Lipomucopolysaccharidose. Eine speicherkrankheit. *Zschr Kinderh* 1968; 103: 285-306.
31. Cantz M, Gehler J, Spranger J. Mucolipidosis I: Increased sialic acid content and deficiency of an a-N-acetylneuraminidase in cultured fibroblasts. *Biochem Biophys Res Commun* 1977; 74: 732-38.
32. Leroy JG, Demars RI. Mutant enzymatic and cytological phenotypes in cultured human fibroblasts. *Science* 1967; 157: 804-06.
33. Hasilik A, Waheed A, von Figura K. Enzymatic phosphorylation of lysosomal enzymes in the presence of UDP-N-acetylglucosamine. Absence of the activity in I-cell fibroblasts. *Biochem Biophys Res Commun* 1981; 98: 761-67.
34. Berman ER, Livni N, Shapira E, Merin S, Levis IS. Congenital corneal clouding with abnormal systemic storage bodies: A new variant of mucolipidosis. *J Pediatr* 1974; 84: 519-26.
35. Bach G, Zeigler M, Schaapt, Kohn G. Mucolipidosis Type IV: Ganglioside sialidase deficiency. *Biochem Biophys Res Commun* 1979; 90: 1341-47.
36. Ockerman PA. A generalized storage disorder resembling Hurler's syndrome. *Lancet* 1967; 2: 239-41.
37. Hultberg B. Properties of a-mannosidase in mannosidosis. *Scand J Clin Lab Invest* 1970; 26: 155-59.
38. Durand P, Borrone C, Della Cella G. A new mucopolysaccharide lipid-storage disease? *Lancet* 1966; 2: 1313-14.
39. Van Hoof F, Hers HG. Mucopolysaccharidosis by absence of a-fucosidase. *Lancet* 1968; 1: 1198.
40. Jenner FA, Pollitt RJ. Large quantities of 2-acetamido-l(β-L-aspartamido)-1, 2-dideoxyglucose in the urine of mentally retarded siblings. *Biochem J* 1967; 103: 48P.
41. Pollitt RJ, Jenner FA, Merskey H. Aspartylglycosaminuria. An inborn error of metabolism associated with mental defect. *Lancet* 1968; 2: 253-55.
42. Aula P, Autio S, Raivio KO, *et al.* "Salla disease" A new lysosomal storage disorder. *Arch Neurol* 1979; 36: 88-94.
43. Sandhoff K, Jatzkewitz H. The chemical pathology of Tay-Sachs disease. *In*: Volk B, Aronson SM, eds. Sphingolipids, sphingolipidosis and allied disorders. New York: Plenum Press, 1972: 305-19.
44. Kaback MM, ed. Tay-Sachs disease: Screening and prevention. Progress in clinical and biological research. Vol. 18. New York: Alan R Liss, 1977.
45. Lowden JA. Quality control for hexosaminidase assays for Tay-Sachs disease carrier identification. *Am J H Genet* 1981; 33:7A.
46. Lowden JA, Callahan JW, Howard F. Hexosaminidases: Multiple component enzymes. *In:* Volk BW, Schneck L, eds. Current trends in sphingolipidoses and allied disorders. New York: Plenum Press, 1975: 313-22.
47. MacLeod PM, Wood S, Jan TE, Applegarth DA, Bolman CL. Progressive cerebellar ataxia, spasticity, psychomotor retardation and hexosaminidase deficiency in a ten year old child: Juvenile Sandhoff disease. *Neurology* 1977; 27: 571-73.

48. Willner JP, Grabowski GA, Gordon RE, Bender AN, Desnick RJ. Chronic GM2 gangliosidosis masquerading as atypical Friedreich ataxia: Clinical, morphologic and biochemical studies of nine cases. *Neurology* 1981; 31: 787-98.

49. Rattazzi MC, McCullough RA, Downing CJ, Kung M-P. Towards enzyme therapy in GM2 gangliosidosis: β-hexosaminidase infusion in normal cats. *Pediat Res* 1979; 13:916-23.

50. Lowden JA. Approaches to the diagnosis and management of infants and children with lysosomal storage disease. *In:* Kaback MM, ed. Genetic issues in pediatrics and perinatology. Chicago: Medical Year Book 1981: 267-305.

51. de Duve C. *In:* Dingle JT, Fell HB, eds. Lysosomes in biology and pathology. Amsterdam:North-Holland, 1969: 3-40.

52. Varki A, Kornfeld S. Purification and characterization of rat liver a-N-acetylglucosaminyl phosphodiesterase. *J Biol Chem* 1981; 256: 9937-43.

53. Terry RD, Weiss M. Studies in Tay-Sachs disease II. Ultrastructure of the cerebrum. *J Neuropath Exp Neurol* 1963; 22: 18-55.

54. Gonatas NK, Terry RD, Winkler R, Korey SR, Gomez CJ, Stein A. A case of juvenile lipidosis: The significance of electron microscopic and biochemical observations of cerebral biopsy. *J Neuropath Exp Neurol* 1963; 22: 557-86.

55. Brady RO, O'Brien JS, Bradley RM, Gal AE. Sphingolipid hydrolases in brain tissue of patients with generalized gangliosidosis. *Biochim Biophys Acta* 1970; 210: 193-95.

56. Callahan JW, Gerrie J. Purification of GM1 ganglioside and ceramide lactose β-galactosidase from rabbit brain. *Biochim Biophys Acta* 1975; 391: 141-53.

57. O'Brien JS, Norden GW, Miller AL, Frost RG, Kelly TE. Ganglioside GM2 N-acetyl-β-D-galactosaminidase; studies in human skin fibroblasts. *Clin Genet* 1977; 11: 171-83.

58. Strasberg PM, Lowden JA. Assay of glucocerebrosidase activity using the natural substrate. *Clin Chim Acta* 1981; 118: 9-20.

59. Deleted in editing.

60. Rao BG, Spence MW. Niemann-Pick disease type D: Lipid analysis and studies on sphingomyelinases. *Ann Neurol* 1977; 1: 385-92.

61. Novak A, Lowden JA, Gravel YL, Wolfe LS. Preparation of radiolabelled GM2 and GA2 gangliosides. *J Lipid Res* 1979; 20: 678-81.

62. Schneider PB, Kennedy EP. Sphingomyelinase in normal human spleens and in spleens from subjects with Niemann-Pick disease. *J Lipid Res* 1967; 8: 202-09.

63. Yedgar S, Gatt S. Effect of Triton X-100 on the hydrolysis of sphingomyelin by sphingomyelinase of rat brain. *Biochemistry* 1976; 15: 2570-73.

64. Conzelmann E, Sandhoff K. AB variant of infantile GM2 gangliosidosis: Deficiency of a factor necessary for stimulation of hexosaminidase A-catalyzed degradation of ganglioside GM2 and glycolipid GA2. *Proc Nat Acad Sci USA* 1978; 75: 3979-83.

65. Novak A, Lowden JA. Kinetics of rat liver GM2 ganglioside β-D-N-acetylhexosaminidase. *Can J Biochem* 1980; 58: 82-8.

66. Jones CS, Davidson DJ, Callahan JW. Complex kinetics of bis(4-methylumbelliferyl) phosphate and hexadecanoyl (nitrophenyl) phosphorylcholine hydrolysis by purified sphingomyelinase in the presence of Triton X-100. *Biochem Biophys Acta* 1982; 701: 261-68.

67. O'Brien JS, Tennant L, Veath ML. Characterization of unusual hexosamin-idase A (HEX A) deficient human mutants. *Am J Hum Genet* 1978; 30: 602-08.
68. Gatt S, Dinur T, Barenholz Y. A fluorometric determination of sphingo-myelinase by use of fluorescent derivatives of sphingomyelin and its application to diagnosis of Niemann-Pick disease. *Clin Chem* 1980; 26: 93-6.
69. Fischer G, Jatzkewitz H. The activator of cerebroside sulfatase. Purifi-cation from human liver and identification as protein. *HS Zeit Physiol Chem* 1975; 356: 605-13.
70. Li S-C, Li Y-T. An activator stimulating the enzymic hydrolysis of sphin-goglycolipids. *J Biol Chem* 1976; 251: 1159-63.
71. Ho MW. Specificity of low molecular weight glycoprotein effector of lipid glycosidase. *FEBS Lett* 1975; 53: 243-47.
72. Shapiro LJ, Aleck KA, Kaback MM, et al. Metachromatic leukodystrophy without arylsulfatase A deficiency. *Pediat Res* 1979; 13: 1179-81.
73. Christomanou H. Niemann-Pick disease type C: Evidence for the de-ficiency of an activating factor stimulating sphingomyelin and gluco-cerebroside degradation. *HS Zeit Physiol Chem* 1980; 361: 1489-1502.
74. Suzuki Y, Suzuki K. Partial deficiency of hexosaminidase component A in juvenile GM_2-gangliosidosis. *Neurology* 1970; 20: 848-51.
75. Harzer K. Enzymic diagnosis in 27 cases with Gaucher's disease. *Clin Chim Acta* 1980; 106: 9-15.
76. Spence MW, Ripley BA, Embil JA, Tibbles JAR. A new variant of Sand-hoff's disease. *Ped Res* 1974; 8: 628-37.
77. Raghavan SS, Topol J, Kolodny EH. Leukocyte β-glucosidase in homo-zygotes and heterozygotes for Gaucher disease. *Am J Hum Genet* 1980; 32: 158-73.
78. Lowden JA, Callahan JW, Gravel RA, Skomorowski MA, Becker L, Groves J. Type 2 GM_1 gangliosidosis with long survival and neuronal ceroid lipofuscinosis. *Neurology* 1981; 31: 719-24.
79. Gravel RA, Lowden JA, Callahan JW, Wolfe LS, Ng Kim NMK. Infantile sialidosis: A phenocopy of GM_1 gangliosidosis distinguished by genetic complementation and urinary oligosaccharides. *Am J Hum Genet* 1979; 31: 669-79.
80. Galjaard H, Hoogeveen A, Verheijen F, et al. Relationship between clinical biochemical and genetic heterogeneity in sialidase deficiency. *Perspec Inherit Metab Dis* 1981; 4: 317-37.
81. Shapiro LJ. Current studies and future direction for carrier detection in lysosomal storage diseases. *In*: Callahan JW, Lowden JA, eds. Lysosomes and lysosomal storage diseases. New York: Raven Press, 1981: 343-55.
82. Lowden JA, Zuker S, Wilensky AJ, Skomorowski MA. Screening for carriers of Tay-Sachs disease: A community project. *Can Med Ass J* 1974; 111: 229-33.
83. Lott IT, Dulaney JT, Milunsky A, Hoefnagel D, Moser HW. Apparent biochemical homozygosity in two obligatory heterozygotes for meta-chromatic leukodystrophy. *J Pediatr* 1976; 89: 438-40.

GENETIC HETEROGENEITY AND COMPLEMENTATION ANALYSIS: GENERAL PRINCIPLES AND STUDIES IN PROPIONICACIDEMIA

R. A. Gravel

Many inborn errors of metabolism show considerable heterogeneity in the age of onset, expression and clinical course. It is important to determine the basis of such variation. At one extreme it may result from grouping more than one disease under the same clinical entity. In other cases mutations at more than one gene may result in the expression of the same clinical disorder. Finally, there may be heterogeneous expression of different allelic mutations. These alternatives complicate the effort to provide accurate diagnosis, treatment, prognosis and genetic counseling. It is becoming evident that accurate diagnosis of genetic disease should depend on detection of the primary lesion, *i.e.* identification of the gene and its alleles.

THE NATURE OF COMPLEMENTATION ANALYSIS

The most useful technique currently available to delineate the genetic basis of a metabolic disorder is the complementation test.[1] It requires bringing together the defective gene of each mutant into the same cell and testing for restoration of function. A normal phenotype should result if each mutant can supply what is deficient in the other. This will be the case for mutations in different genes (intergenic complementation) and, in rare instances, for different mutations in the same gene (intragenic complementation). Retention of the mutant phenotype is *a priori* evidence that the mutations occurred in the same genetic locus, since neither parent genome can correct for the defect of the other. These points apply to recessive mutations alone, since a dominant mutation would not permit the expression of the normal allele in a complementation test.

Most complementation studies of human metabolic disorders have been done in fibroblast heterokaryons. These are multinucleate cells resulting from the fusion of different mutant strains in the presence of inactivated Sendai virus[2] or polyethylene glycol (PEG).[3] The result of cell fusion is to produce a mixed cell population containing, in addition to heterokaryons, unfused parental cells and multinucleate homokaryons containing nuclei of only one parent. True heterokaryons are generally present in from 5–20% of the total

TABLE 1
Complementation Analysis

Disorder	Complementation Groups	Reference
Galactosemia	Intragenic	27
Maple syrup urine disease	2	6,28
Methylmalonicacidemia	3	12
Combined methylmalonicacidemia and homocystinuria	2	12,29
Propionicacidemia	2	10,18
Biotin-responsive multiple carboxylase deficiency	1	18
Citrullinemia	1	30
Arginosuccinicaciduria	Intragenic	30
β-galactosidase deficiency (GM$_1$ gangliosidosis, sialidosis)	2	8,11,31
Neuraminidase deficiency (mucolipidosis I, sialidosis)	2?	32,33
I-cell disease/pseudo-Hurler polydystrophy	2	9
Tay-Sachs disease/Sandhoff disease	2	34,35
Niemann-Pick disease	2	36
Gaucher disease	1	*
Sulfatase deficiencies (multiple sulfatase, metachromatic leukodystrophy, Hunter, Sanfilippo A)	4	38,39,40
Cockayne syndrome	2	41
Xeroderma pigmentosum	7	42,43

*Gravel RA, Leung A. Complementation analysis of Gaucher disease by single cell microtechniques. Unpublished observations.

cell population, although multinucleate cells can now be enriched over unfused cells by gravity sedimentation.[4] It is only these cells which are competent to demonstrate the presence or absence of complementation. They do not divide in culture but are metabolically active for up to two weeks. In rare instances, nuclear fusion results in the formation of true cell hybrids which do grow in culture and which are beginning to be used in complementation experiments.[5]

During the past decade several disorders from different areas of metabolism have been examined by complementation (Table 1). The major problem is the development of a method for detecting correction of the defect. It has often been possible to simply assay extracts of the fused cell population for the relevant enzyme.[6,7] In other cases more novel detection methods have been required, such as monitoring an enzyme or pathway by a histochemical[8,9] or autoradiographic method.[10] At a more exotic extreme, individual hetero-karyons have been extracted and assayed directly for enzyme activity.[11]

One of the major achievements of complementation analysis is the de-lineation of the gene system involved in the expression of a genetic disorder or group of such diseases. As shown in Table 1, several disorders have been found in which from two to as many as seven genes are involved. In some in-stances these studies have provided the initial evidence of the involvement of several steps of a pathway as possible sites of mutation in a metabolic dis-order.[12,13]

THE GENE SYSTEM IN PROPIONICACIDEMIA

In this presentation, I will describe the progress our laboratory and others have made in characterizing the genetic system involved in propionicacidemia through complementation analysis.[14] Our studies of this disorder which re-sults from a deficiency of the activity of the biotin-dependent enzyme, propionyl-CoA carboxylase (PCC), indicate that a multigene system is involved in the expression of PCC activity, including two structural genes encoding the polypeptides of PCC and a third gene required for activation of the enzyme.

The disease, propionicacidemia, is clinically heterogeneous. Some affected infants die in the newborn period; others survive but may be mentally re-tarded.[14] In rare instances the disorder is virtually benign, despite the indi-vidual's nearly complete deficiency of enzyme activity.[15]

We first sought to determine if there were a genetic basis to the clinical heterogeneity in propionicacidemia by examining complementation between fibroblast cultures from affected patients. We reasoned that the disorder might be related to mutations in different genetic loci required for the expression of the enzyme (Figure 1). At least one gene would be the structural gene encoding propionyl-CoA carboxylase. Another gene would be required to catalyze the covalent linkage of the biotin cofactor of the enzyme. This gene would code for a holocarboxylase synthetase. Finally, a gene would be required to account for biotin transport into cells. If any of these proteins were multi-meric and contained additional polypeptides, then additional genes might be expected. Also, if biotin were metabolized before it activated the apo-carboxylase, then additional genes would be required for these steps.

COMPLEMENTATION STUDIES

To examine complementation we used a method to detect the conversion of ^{14}C-propionate to ^{14}C-labelled protein, a sequence that would require a functional propionyl-CoA carboxylase and metabolism through the Krebs cycle. By using autoradiography to detect ^{14}C-protein we readily detected the presence or absence of complementation in multinucleate cells.[10] Direct

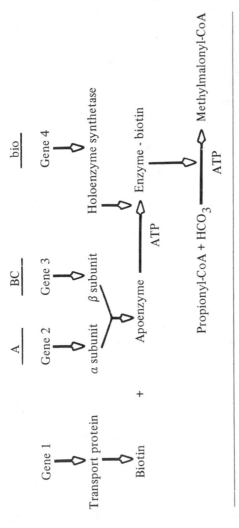

Fig. 1. Gene requirements for the activation of propionyl-CoA carboxylase. Gene 1 is proposed to encode a protein required for the transport into cells of the cofactor biotin. Gene 2 (complementation group *pccA*) and gene 3 (*pccBC*) are thought to encode the α and β chains of the apocarboxylase, respectively. Gene 4 (bio) is proposed to encode the holo-enzyme synthetase required for the activation of the apocarboxylase by biotin.

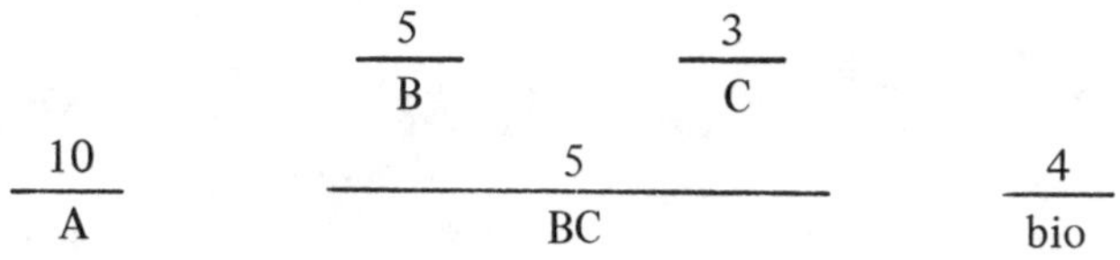

$$\begin{array}{ccc}
\dfrac{5}{B} & \dfrac{3}{C} & \\
\dfrac{10}{A} & \dfrac{5}{BC} & \dfrac{4}{bio}
\end{array}$$

Fig. 2. Complementation map of propionyl-CoA carboxylase deficiency. Letters refer to *pcc* complementation groups. The number of mutants mapped to each group is indicated on the solid line. The map is read as follows: Mutant strains placed within a group fail to complement each other and are placed over a solid line to define that group. Strains placed in different groups over nonoverlapping lines (*pccA* vs. *pccB, pccB* vs. *pccC)* complement each other, while strains placed in different groups over overlapping lines fail to complement each other (*pccB* or *pccC* vs. *pccBC*). Note that mutant strains in the same group behave identically in complementation tests.

---

radioactive counting has also been used on complete fused cell populations[16] and on isolated multinucleate cells.[17]

From these results we produced a complementation map (Figure 2),[18] which showed two principle complementation groups, *pccA* and *pccBC*, accounting for propionyl-CoA carboxylase deficiency. It appeared that two genes were involved in the expression of the enzyme, and that their interaction required protein synthesis,[16] as would be expected for different polypeptides of a multimeric enzyme.

POLYPEPTIDE STRUCTURE OF PROPIONYL-CoA CARBOXYLASE

The enzyme from human liver has recently been purified by our laboratory[19] and by Kalousek *et al*[20] and proved to contain two different polypeptides, a biotin-containing α chain of 75,000 daltons and a β chain of 60,000 daltons. The enzyme appeared to occur as an $\alpha_4\beta_4$ octomer with a native molecular weight of 540,000 daltons and a sedimentation coefficient of 17.4 S. The occurrence of two polypeptides correlated well with the presence of two *pcc* complementation groups. The assignment of the two polypeptides to the putative structural genes will be discussed later. Rosenberg and his colleagues have used mRNA translation *in vitro* to show that the α and β chains (rat liver) are products of independent mRNA species, a result consistent with a requirement for two structural genes.[21]

IDENTIFICATION OF GENE PRODUCTS

In order to assign specific gene products to the *pccA* and *pccBC* complementation groups, we have examined ^{35}S-methionine-labelled PCC immunoprecipitated from mutant fibroblasts.[22]*Our initial results showed that most mutants from *pccBC,* or its subgroups, failed to synthesize any antigenically active β chain but did appear to have a normal, biotin-bound α chain. However, mutants of the *pccA* group had neither α nor β chains by anti-PCC precipitation. In the absence of detecting structurally altered polypeptides, we can only speculate about the gene assignments at this time. We suggest that *pccBC* encodes the β chains and, accordingly, synthesizes a detectable α, but not a β chain. Consequently the *pccA* group would appear to encode the α chain. This might explain the absence of α chains in *pccA* mutants. On the other hand, the inability to detect any β chain in *pccA* mutants might be caused by a lack of antigenecity of free β chains, or by the presence of β chains in severely modified nonantigenic multimers.

INTRAGENIC COMPLEMENTATION

An unexpected finding in the complementation analysis was the intragroup complementation among *pccBC* mutants. These subgroups, *pccB* and *pccC* (Figure 2), are made up of mutants which complement each other but which fail to complement those assigned to *pccBC*.[14,17] The occurrence of intragenic complementation provides compelling evidence, albeit indirect, that the complementation group represents a structural gene and that the gene product associates as a multimer. In the case of *pccBC,* the two subgroups are well defined and distinct. Possibly, they correspond to different sets of clustered mutations along the polypeptide chain, perhaps corresponding to different functional centers. For example, if the β chain (the putative gene product) has more than one active site, mutations in each of these sites might be expected to complement since the multimer contains four β chains. Thus a hybrid enzyme would have functional forms of the two active sites on different β chains of the same multimer.

BIOTIN–RESPONSIVE MULTIPLE CARBOXYLASE DEFICIENCY

One additional complementation group has been identified leading to PCC deficiency, the *bio* group (Figure 2). Patients from this group have biotin-responsive multiple carboxylase deficiency, a disorder in which at least three carboxylase activities are deficient, including those of PCC, pyruvate carboxylase, and β-methylcrotonyl-CoA carboxylase, all biotin-dependent mitochondrial enzymes.[17] Recently it has been suggested that acetyl-CoA carboxylase, a biotin-dependent enzyme occurring in the cytoplasma, is also defective in this disorder.[23]

*Gravel RA, Lam KF, Lam AM. Unpublished observations.

The multiple carboxylase defect appears to be a secondary manifestation caused by an inability to biotinylate the carboxylases. We[24] and Sweetman and his colleagues[25] have demonstrated a defect of holocarboxylase synthetase in lymphoblasts and fibroblasts of affected patients. The latter authors showed that the Km of the enzyme for biotin was increased in their patient.[25]

Finally, a variant of the biotin-responsive disorder shows the multiple carboxylase defect in tissues but not in fibroblasts.[26] It has been suggested that these patients have a failure of biotin transport, possibly restricted to the site of intestinal absorption of biotin.[26] These mutants cannot be examined by complementation in the absence of cell culture manifestations of the disorder.

These studies have shown that complementation analysis can be used to reveal the gene system involved in a metabolic disorder. In combination with biochemical analysis, it has been possible to investigate the nature of the gene products required for the expression of the normal enzyme. These studies illustrate the need for gene-level identification of human metabolic disease to ensure the accurate diagnosis of clinically heterogeneous disorders.

ACKNOWLEDGMENTS

I should like to acknowledge the contributions, advice and collaboration of K. F. Lam, M. Saunders, Y. E. Hsia, L. E. Rosenberg and L. Sweetman. These studies were supported by grant MA5698 from the Medical Research Council of Canada.

REFERENCES

1. Fincham JRS. Genetic complementation. New York: Benjamin, 1966.
2. Klebe RJ, Chan TR, Ruddle FH. Controlled production of proliferating somatic cell hybrids. *J Cell Biol* 1970; 45: 74-82.
3. Davidson RL, Gerald PS. Induction of mammalian somatic cell hybridization by polyethylene glycol. *In*: Prescott PM, ed. Methods in cell biology. Vol. 15. New York: Academic Press, 1977: 325-38.
4. Hohman LK, Shows TB. Complementation of genetic disease: A velocity sedimentation procedure for the enrichment of heterokaryons. *Somat Cell Genet* 1979; 5: 1013-29.
5. Chang DL, Davidson RG. Complementation of arylsulfatase A in somatic hybrids of metachromatic leukodystrophy and multiple sulfatase deficiency disorder fibroblasts. *Proc Natl Acad Sci USA* 1980; 77: 6166-70.
6. Singh S, Willers I, Goede HW. Heterogeneity in maple syrup urine disease: Aspects of cofactor requirement and complementation in cultured fibroblasts. *Clin Genet* 1977; 11: 277-84.

7. Horwitz, AL. Genetic complementation studies of multiple sulfatase deficiency. *Proc Natl Acad Sci USA* 1979; 76: 6496-99.

8. Gravel RA, Lowden JA, Callahan JW, Wolfe LS, Ng Yin Kim NMK. Infantile sialidosis: A phenocopy of type I GM_1 gangliosidosis distinguished by genetic complementation and urinary oligosaccharides. *Amer J Hum Genet* 1979; 31: 669-79.

9. Gravel RA, Gravel YL, Miller AL, Lowden JA. Genetic complementation analysis of I-cell disease and pseudo-Hurler polydystrophy. *In*: Callahan JW, Lowden JA, eds. Lysosomes and lysosomal storage diseases. New York: Raven, 1980.

10. Gravel RA, Lam KF, Scully KJ, Hsia YE. Genetic complementation of propionyl-CoA carboxylase deficiency in cultured human fibroblasts. *Amer J Hum Genet* 1977; 29: 378-88.

11. Galjaard H, Hoogeveen A, Keijzer W, *et al.* Genetic heterogeneity in GM_1 gangliosidosis. *Nature* 1975; 257: 60-2.

12. Gravel RA, Mahoney MJ, Ruddle FH, Rosenberg LE. Genetic complementation in heterokaryons of human fibroblasts defective in cobalamin metabolism. *Proc Natl Acad Sci USA* 1975; 72: 3181-85.

13. Mellman I, Willard HF, Youngdahl-Turner P, Rosenberg LE, Cobalamin enzyme synthesis in normal and mutant human fibroblasts. *J Biol Chem* 1979; 254: 11847-53.

14. Wolf B, Hsia YE, Sweetman L, Gravel RA, Harris DJ, Nyhan WL. Propionicacidemia: A clinical update. *J Pediatr* 1981; 99,6: 835-46.

15. Wolf B, Paulsen EP, Hsia YE. A symptomatic propionyl-CoA carboxylase deficiency in a 13 year old girl. *J Pediatr* 1979; 95: 563-65.

16. Wolf B, Willard HF, Rosenberg LE. Kinetic analysis of genetic complementation in heterokaryons of propionyl-CoA carboxylase deficient human fibroblasts. *Amer J Hum Genet* 1980; 32: 16-25.

17. Gravel RA, Leung A, Saunders M, Hosli P. Complementation analysis using whole cell microtechniques. *Proc Natl Acad Sci USA* 1979; 76: 6520-24.

18. Saunders M, Sweetman L, Robinson B, Roth K, Cohn R, Gravel RA. Biotin responsive organicaciduria. Multiple carboxylase defects and complementation studies with propionicacidemia in cultured fibroblasts. *J Clin Invest* 1979; 64: 1695-1702.

19. Gravel RA, Lam KF, Mahuran D, Kronis A. Purification of human liver propionyl-CoA Carboxylase by carbon tetrachloride extraction and monomeric avidin affinity chromatography. *Arch Biochem Biophys* 1980; 201: 669-73.

20. Kalousek F, Darigo MD, Rosenberg LE, Isolation and characterization of propionyl-CoA carboxylase from normal human liver. *J Biol Chem* 1980; 255: 60-5.

21. Kraus JP, Kalousek F, Rosenberg LE. Cell free translation and processing of the precursors of propionyl-CoA carboxylase subunits. *Amer J Hum Genet.* 1981; 33: 47A.

22. Deleted in editing.

23. Feldman GL, Wolf B. Deficient acetyl-CoA carboxylase activity in multiple carboxylase deficiency. *Clin Chim Acta* 1981; 111: 147-51.

24. Saunders ME, Sherwood WG, Duthie M, Surh L, Gravel RA. Evidence for a defect of holocarboxylase synthetase activity in cultured lymphoblasts from a patient with biotin-responsive multiple carboxylase deficiency. *Amer J Hum Genet.* In press.

25. Burri BS, Sweetman L, Nyhan WL. Mutant holocarboxylase synthetase.

Evidence for the enzyme defect in early infantile biotin-responsive multiple carboxylase deficiency. *J Clin Invest* 1981; 68: 1491-95.

26. Munnich A, Saudubray JM, Ogier H, *et al*. Deficit multiple des carboxylases. Une maladie metabolique vitamino-dependante curable par la biotine. *Arch Fr Pediatr* 1981; 38: 83-90.

27. Nadler HL, Chacko CM, Rachmeler M. Interallelic complementation in prosphate uridyl transferase activity. *Proc Natl Acad Sci USA* 1970; 67: 976-82.

28. Lyons LB, Cox RP, Dancis J. Complementation analysis of maple syrup urine disease in heterokaryons derived from human diploid fibroblasts. *Nature* 1973; 243: 533-35.

29. Willard HF, Mellman IS, Rosenberg LE. Genetic complementation among inherited deficiencies of methylmalonyl-CoA mutase activity: Evidence for a new class of human cobalamin mutant. *Amer J Hum Genet* 1978; 30: 1-13.

30. Cathelineau L, Pham Dinh D, Briand P, Kamoun P. Studies on complementation in arginosuccinate synthetase and arginosuccinate lyase deficiencies in human fibroblasts. *Hum Genet* 1981; 57: 282-84.

31. Hoeksema HL, Van Diggelin OP, Galjaard H. Intergenic complementation after fusion of fibroblasts from different patients with β-galactosidase deficiency. *Biochim Biophys Acta* 1979; 566: 72-9.

32. Kato T, Okada S, Yutaka T, *et al*. β-galactosidase deficient-type mucolipidosis: A complementation study of neuraminidase in somatic cell hybrids. *Biochim Biophys Res Commun* 1979; 91: 114-17.

33. Hoogeveen AT, Verheijen FW, d'Azzo A, Galjaard H. Genetic heterogeneity in human neuraminidase deficiency. *Nature* 1980; 285: 500-02.

34. Gaaljaard H, Hoogeveen A, de Wit-Verbeek HA, *et al*. Tay-Sachs and Sandhoff disease: Intergenic complementation after somatic cell hybridization. *Exp Cell Res* 1974; 87: 444-48.

35. Rattazzi MC, Brown JA, Davidson RG, Shows TB. Studies on complementation of β-hexosaminidase deficiency in human GM_2 gangliosidosis. *Amer J Hum Genet* 1976; 28: 143-54.

36. Besley GTN, Hoogeboom AJM, Hoogeveen A, Kleijer WJ, Galjaard H. Somatic cell hybridization studies showing different gene mutations in Niemann-Pick variants. *Hum Genet* 1980; 54: 409-12.

37. Deleted in editing.

38. Eisenberg LR, Migeon BR. Enrichment of human heterokaryons by ficoll gradient for complementation analysis of iduronate sulfatase deficiency. *Somat Cell Genet* 1979; 5: 1079-89.

39. Howritz AL. Genetic complementation studies of multiple sulfatase deficiency. *Proc Natl Acad Sci USA* 1979; 76: 6496-99.

40. Chang PL, Davidson RG. Complementation of arylsulfatase A in somatic hybrids of metachromatic leukodystrophy and multiple sulfatase deficiency disorder fibroblasts. *Proc Natl Acad Sci USA* 1980; 77: 6166-70.

41. Tanaka K, Kawai K, Kumahara Y, Ikenaga M, Okada Y. Genetic complementation groups in Cockayne syndrome. *Somat Cell Genet* 1981; 7: 445-55.

42. Kraemer KH, de Weerd-Kastelein EA, Robbins JH, *et al*. Five complementation groups in xeroderma pigmentosum. *Mutation Res* 1975; 33: 327-40.

43. Hashem N, Bootsma D, Keijzer W, *et al*. Clinical characteristics, DNA repair and complementation groups in xeroderma pigmentosum patients from Egypt. *Cancer Res* 1980; 40: 13-8.

DE NOVO STRUCTURAL REARRANGEMENTS: IMPLICATIONS FOR PRENATAL DIAGNOSIS

Dorothy Warburton

THE NATURE OF THE PROBLEM

De novo structural rearrangements of chromosomes (those not inherited from a parent) which might be found at prenatal diagnosis fall into three main categories: (a) apparently unbalanced rearrangements, (b) small extra "marker" chromosomes usually of undetermined origin, and (c) apparently balanced rearrangements (chiefly Robertsonian and reciprocal translocations) and inversions.

In the first category, the diagnostic problem is the least difficult. The presence of missing or additional autosomal material is usually considered sufficient to predict serious physical or mental abnormality, even though the chromosomal segments involved may not be identifiable when the parents have normal karyotypes. Of course, it is important to demonstrate using appropriate staining techniques that the chromosomal material involved is not one of the blocks of heterochromatin known to be polymorphic in size without obvious effect on the phenotypes. Sex chromosome anomalies and some ring chromosomes will still present problems in prognosis, but in general the situation is less ambiguous than in the other types of *de novo* rearrangements to be discussed. All available evidence suggests that a chromosomal deletion or addition large enough to be seen cytologically does not occur without serious consequences. The delineation of many new chromosomal syndromes is also providing more specific information on phenotype when the chromosomal segment involved can be identified. However, it might be noted that high resolution banding techniques may soon make it possible to recognize changes so small that we have no basis for predicting their phenotypic effects.

In the second category, the supernumerary small marker chromosomes, prognosis is very difficult. It seems clear from the literature that in many such cases, both familial and non-familial, the additional material has no apparent phenotypic effect.[1,2] However, such additional markers have also been associated with malformation syndromes such as the "cat-eye" syndrome,[3] and with rather non-specific behavioral and neurological problems.[4] Specific identification of the origin of the marker is sometimes possible,[5] sometimes impossible. The problem is compounded by the fact that these supernumerary markers are frequently found in a mosaic state.

The third category, *de novo* presumptive balanced rearrangements, has been the subject of a good deal of speculation. It was first suggested by Breg *et al*,[6] and later by Jacobs,[7] that such rearrangements might not be without phenotypic effect, because they appeared to occur more often than expected in populations of the mentally retarded. In the absence of phenotypically normal relatives carrying the same rearrangement, the cytogeneticist must first of all try to decide, within the limits of available techniques, if the rearrangement is truly balanced. Even then, it is clear that small deletions or duplications cannot be ruled out. Also, it is possible that in the absence of changes in the amount of chromosomal material, rearrangements may affect gene function. Such a "position effect", known in other organisms, has been made particularly credible by our new knowledge of the actions of transposable elements, and their probable presence in the human genome.

Although these *de novo* rearrangements are rare, there can be few prenatal cytogenetic laboratories in the country which have not encountered one of these difficult diagnostic problems. There is clearly a need for empirical data concerning the risk of abnormality in these cases. In this paper I will try to present what data are available for providing risk estimates at prenatal diagnosis, including new data gathered from a mail survey of U.S. cytogenetic laboratories. However, there is still insufficient data for a satisfactory answer.

SOURCES OF DATA

Two kinds of information will be useful in evaluating the risk of serious abnormality among those with balanced *de novo* rearrangements and supernumerary markers. The first is the relative frequencies of such abnormalities in various populations, including the retarded or malformed. If the frequency is higher in such groups than in random samples of newborns, then an increased risk for abnormalities in patients with these rearrangements is indicated. While some idea of the magnitude of the increased risk can also be deduced, this kind of data does not permit a direct estimate of the relative risk.

The other kind of information needed is the phenotypic outcome of cases ascertained through random population surveys, such as newborns, or fetuses studied through amniocentesis. The amniocentesis data are not representative of births in general in several ways, particularly with respect to maternal age and social class, but this should not seriously effect the rate of abnormality found in fetuses with *de novo* rearrangements. The newborn data are more seriously biased towards normality, because late fetal deaths, stillbirths and early neonatal deaths are all absent or underrepresented.

As part of an international study group scheduled to report to the Sixth International Congress of Human Genetics in Jerusalem, I was asked to collect data from the U.S. on the outcome of pregnancies diagnosed at amniocentesis with balanced *de novo* rearrangements, and supernumerary markers. I sent a short questionnaire to about 200 prenatal diagnosis centers listed by the National Clearing House for Genetic Diseases. The response was excellent. I received about 80 replies from laboratories in 40 states, plus a number of letters explaining that a particular center referred its laboratory work elsewhere. I requested information on the number of cases of each type of abnormality seen over a period of at least one year or 100 amniocenteses, the cytogenetic diagnosis of each case, its outcome, and the source and duration of follow-up. Information on 76,952 prenatal diagnoses was available at last analysis.

The data on newborns, spontaneous abortions and populations of the retarded are derived from the literature using primarily the same sources as Jacobs[7-9] and Funderburk *et al.*[10] My figures for the mentally retarded differ somewhat from those previously reported by Funderburk *et al* partly due to errors in their summary, and partly due to updating with subsequent studies.

FREQUENCY OF *DE NOVO* BALANCED REARRANGEMENTS IN VARIOUS POPULATIONS

Table 1 shows the estimates of the frequency of balanced rearrangements in several kinds of population surveys. In the 59,452 newborns studied the incidence was 0.19%. It was only slightly higher than this in spontaneous abortions and in the small number of karyotyped consecutive stillbirths and neonatal deaths. The incidence is significantly increased to 0.44% ($p<.0001$) in the mentally retarded. The increase appears to be almost entirely among the non-Robertsonian rearrangements.

Table 2 shows that it is the *de novo* rather than the familial balanced rearrangements which show changes in frequency. For each population, the proportion of *de novo* rearrangements among all balanced rearrangements was calculated from those cases in which both parents were studied,* and then the total frequency of *de novo* rearrangements estimated. While 20% of balanced rearrangements in the newborn were *de novo,* this frequency was over 50% of those studied in the retarded. The estimated overall incidence of *de novo* rearrangements in the retarded is almost seven times that in the newborn. The greatest increase is seen in non-Robertsonian *de novo* rearrangements. However, there is a suggestion of an increase in Robertsonian rearrangements also, which is not conclusive because of the small number of such cases.

---

*This was done separately for Robertsonian and non-Robertsonian rearrangements for newborns and the mentally retarded.

TABLE 1
Frequency of Balanced Rearrangements in Various Populations

Population	Number karyotyped	Balanced Rearrangements					
		Rob.		Others		Total	
		n	%	n	%	n	%
Newborn[9]	59,452	52	0.087	61	0.103	113	0.190
Stillbirth & neonatal deaths	653	1	0.153	1	0.153	2	0.306
Spontaneous abortions	6,462	5	0.077	12	0.186	17	0.263
Mentally[9,10,13-34] retarded	12,629	14	0.111	42	0.332	56	0.443

TABLE 2
Frequency of *De Novo* Rearrangements in Various Populations

Population	No. karyo.	No. bal. rearrang.	No. with parents studied	% de novo	Est. rate de novo rearrangements %		
					Rob.	Others	Total
Newborn	59,452	113	91	19.8	.0087	.0283	.0370
Spontaneous abortions	6,462	17	15	26.7	–	–	.0702
Mentally retarded	12,629	56	33	54.5	.0185	.2232	.2417
Amniocentesis	76,952				.0117	.0442	.0559

There are problems in interpreting these differences in frequency because of the differences in technique used in these surveys. Banding was used for most of the spontaneous abortions, some of the mentally retarded and few of the newborns. However, the difference in frequency between the newborn and the mentally retarded is far too great for a technical explanation.

We can conclude then that an apparently balanced *de novo* rearrangement substantially increases the risk of mental retardation. Some of the mentally retarded populations had other anomalies as well, such as low birth weight or congenital anomalies, but one cannot estimate these risks separately. The risk of spontaneous abortions seems only slightly raised, if at all, which is somewhat surprising in view of the association of most types of chromosome imbalance with a high probability of abortion. Perhaps this is an indication that any structural change is really very small, and that functional changes may be more important.

As indicated in Table 2 the reported incidence of *de novo* balanced rearrangements in my amniocentesis survey is .056, an insignificant increase over that in the newborn, most likely reflecting two sources of bias: (a) There may have been a tendency for labs to return the questionnaire only if they had a case (although 37 of 77 replies reported no *de novo* balanced rearrangements), and (b) all the amniotic fluid cases were banded and most were recent. As shown in Table 3, there does not seem to be any difference in the proportion of balanced Robertsonian, reciprocal translocations or inversions in the series. There is a significant increase in the rate of *de novo* unbalanced rearrangements in the amniocentesis series compared to the newborn series. This may reflect a substantial death rate of these cases in the late fetal and neonatal period. It is also true that unbalanced rearrangements are more likely to be missed in unbanded chromosome preparations (with the exception of the Robertsonian translocation, which show about the same frequency in newborns and amniocentesis). The much higher frequency of cases with supernumerary chromosomes at amniocentesis has no obvious explanation, unless this is associated with parental age.

FREQUENCY OF ABNORMALITY AMONG CASES ASCERTAINED AS NEWBORNS

Table 4 shows the limited amount of data available on the phenotype of 16 cases with a *de novo* rearrangement and five cases with a supernumerary marker ascertained in the newborn studies. Only one serious abnormality was reported in a case with a reciprocal translocation. This case, one of Jacobs, was originally reported to be normal,[11] but in a later paper was described as having developed a severe myoclonic epilepsy with death at age three and a half years.[7] Two cases of congenitally dislocated hip, a case of "asphyxia" at birth and a twin described as "dysmature" were not counted as abnormal since all did well subsequently.

TABLE 3
Comparison of Amniocentesis Cases with Newborn

			Balanced *de novo*			Unbalanced *de novo*		
		No. karyo.	Rob.	Non Rob.	Inv.	Rob.	Others	+Mar.*
Amnio cases	no.	76,952	9	30	4	4	14	15
	%		.012	.039	.005	.005	.018	.019
Newborn	no.	59,452						
(est.)	%		.009	.026	.002	.005	.007	.005

*Only non-mosaic cases.

TABLE 4
Phenotypic Data from Newborn Studies

	Total with known outcome	Nl.	Abnl.
De novo rearrangements			
Robertsonian	4	4	0
Others	12	11	1*
Total	16	15**	1*
			(6.3±6.1)
De novo +Mar.	5	5	0

*Normal at birth; myoclonic epilepsy, died 3½ years.
**Two with congenital dislocated hip.

In most cases there is no record of the newborns having been followed after birth, so that any anomalies not obvious at birth, or any developmental delay, would not have been detected.

In the 4,342 seven year olds karyotyped as part of the Collaborative Perinatal Project, there were three cases detected with demonstrated *de novo* balanced reciprocal translocations. One was said to have a reading disability (but a seven year IQ of 134), one had a club foot, and one had a seven year IQ of 85 and an abnormal skull shape.[12] While this again suggests an increase in the rate of dysfunction, the data are too few for any conclusions.

TABLE 5
Outcome of Amniocentesis Cases with Balanced *De Novo* Rearrangement

	No.	Elect. AB	LB	Nl	Abnl	?
Inversion	4	1	3	1	2*+	1
Robertsonian translocation	9	0	9	9	0	0
Reciprocal translocation	30	7	23	27	1†	2
Total	43	8	35	37	3 (7.4%±4.2)	3

*IUGR, cardiac defects, seizures, severe MR.
+Facial clefting.
† Bilateral renal agenesis.

FREQUENCY OF ABNORMALITY AMONG CASES ASCERTAINED AT AMNIOCENTESIS

Table 5 shows the information available on outcome for cases with a balanced *de novo* rearrangement, ascertained at amniocentesis. Of the 43 non-mosaic cases reported eight were electively terminated. Six of these cases had fetal autopsy. Only seven of the live born cases had been followed for at least six months after birth. We can say almost nothing then about the risk of mental retardation, and the abnormalities which could have been recognized are only those visible at birth or induced abortion. In a future project we will try to collect later follow-up data on the live born infants with these chromosome anomalies.

Three anomalies were reported out of 40 cases with known outcome. Two were in induced abortions and one in a live birth. All were severe defects, as shown in the table. It is interesting that two of the anomalies occurred in cases diagnosed as having balanced inversions rather than translocations. If not just fortuitous, this might reflect the difficulties of distinguishing an inversion (particularly of a small chromosome) from an unbalanced rearrangement involving material from another chromosome.

The overall risk figure for severe anomalies is 3/41 or 7.4%, with 95% confidence limits 0–16%. Thus, it is not possible with a sample this size to rule out a risk as low as the usual rate of severe anomalies at birth (1–2%). However, the risk estimate does agree with that obtained from the live born data (6.3%).

TABLE 6

Outcome of Amniocentesis Cases with Unbalanced *De Novo* Rearrangements

	No.	Elect. term.	LB	Nl	Abnl	?
Robertsonian translocation	4	4	0	0	4 b,e,i,j	0
Others	14	12	2	3	6 a,c,d,f,g,h	5
Total	18	16	2	3	10	5

a Absence of ovarian follicles.
b Malformed ears, cleft palate, ulnar skin tag, enlarged right kidney.
c IUGR, microcephaly.
d Congenital heart defect (coarctation).
e Lumbar myelomingocele, polydactyly, abnormal skull.
f Scoliosis, valgus deformity, hypospadius, cystic hygroma, dysplastic kidney, low set ears.
g Absent flexion creases, malformed ears, overriding fingers.
h Edema, cystic hygroma, digital hypoplasia, coarctation of aorta, hypoplastic lungs, ascites.
i Multiple congenital anomalies consistent with trisomy 13.
j Hexadactyly, occipital sac.

For comparison, Table 6 shows the results for cases of unbalanced *de novo* rearrangements. As expected the risk of anomalies is high, 10/13 or 77%. The surprising thing here is the three cases with no demonstrable anomalies. Two were induced abortions in which no obvious anomalies could be seen, and one was a live born child reported to have an interstitial deletion.

In Table 7 the results of cases with supernumerary chromosomes are displayed. This category of anomaly posed problems in classification, since the markers differed in morphology, and some were mosaic. We include as mosaics only cases where the marker was found in more than one flask or clone, or where it was confirmed in studies of the fetus. Twenty familial cases were reported; interestingly, seven of these were mosaic. The marker chromosomes must have been inherited from a parent, but lost in many of the fetal cells. None of the 13 familial cases of known outcome were abnormal, and all but one involved a satellited marker.

The non-familial cases were grouped into mosaics and non-mosaics, and into morphological groups: satellited (on at least one end), non-satellited, and minute. The total rate of abnormality for the non-familial cases was 6/31 or 19.4% with confidence limits from 5.2% to 33.6%. Thus, there is a significantly increased rate of abnormality for this group. As in the previous table, many cases were induced abortions and follow-up was usually very short,

TABLE 7
Summary of Outcome of Cases with Supernumerary Markers

	Elect. AB	Spont. AB	LB	?	Nl	Abnl	(%)	?
Familial	5	0	15	0	13	0	0.0	7
Non-familial								
mosaic	11	0	10	0	14	4 e,f,g,h	(22.2)	3
non-mosaic	12	1	2	0	11 a,b	2 c,d	(15.4)	2
satellited	10	0	4	0	10 a,b	1 e	(9.1)	3
non-satellited	13	1	8	0	15	5 c,d,f,g,h	(25.0)	2
total	23	1	12	0	25	6	(19.4)	5
? Familial								
mosaic non-satellited	2	1	3		3	2 i,j		1
non-mosaic satellited	1			1				1
non-satellited								1
Non-familial + ? familial								
satellited	11	0	4		10	1	(9.1)±8.0	4
non-satellited	15	2	11	1	18	7	(28.0)±10.0	4

a Mild malformations, low set ears, clinodactyly, broad chest.
b Minor dysmorphic features, hypertelorism, low set ears, micrognathia and anti mongoloid slant.
c Abnormal appearing fetus, two umbilical vessels, micrognathia, abnormal facies, simian crease.
d 47,XY,+21, microcephalic, mongoloid slant, low set ears.
e Grossly normal at birth, at two years microcephaly < third percentile I.Q. 100.
f Malformed ears, simian crease, abnormal facies, imperforate anus.
g Fetal malformations seen.
h Diaphragmatic hernia, severe lung hypoplasia.
i Multiple anomalies, died within minutes of delivery.
j IUGR, microcephaly, micrognathia.

so that only obvious major anomalies will be reported. Two fetuses said to have had minor dysmorphic features or minor malformations were not included as abnormal because of the difficulties in all but very experienced hands of recognizing dysmorphic features in a 20 week fetus. One fetus with a non-mosaic, non-satellited fragment in amniotic fluid culture inexplicably showed trisomy 21 on culture of fetal tissue after abortion.

There is not any strong suggestion of a difference in the rate of abnormality between mosaic and non-mosaic cases, but there is a suggestion of a lower rate of abnormalities for satellited markers than for unsatellited markers. If the cases of unknown origin are added to the non-familial cases, the rates of abnormalty are 9.1±8.0 for satellited cases, and 28.0±10.0 for the non-satellited cases. The difference is not statistically significant.

CONCLUSION

For apparently balanced de novo rearrangements, the data on population surveys indicates an increased rate of mental retardation, which might approach seven times the usual rate in newborns. The data on the outcome of live born cases, and cases ascertained at amniocentesis, agree in showing a small increase (2–3X) in the risk for major malformations. However, these data are too few to rule out a risk as low as that in the general population of newborns. The upper limit of risk is about 15% for major malformations.

For supernumerary marker chromosomes there seems to be a significant increase in the risk of malformations for non-familial cases; the upper limit of risk is about 33%. The available data do not allow any estimate of the risk for mental retardation or the kinds of behavioral and neurological problems which have been reported with some specific markers.[4]

More data is required to enable more useful estimates of risk in these difficult situations. A current attempt to pool European and Canadian data with that from the United States and to improve follow-up of the patients should be helpful in this regard.

ACKNOWLEDGMENTS

The cooperation of all the laboratories which provided data for the amniocentesis survey is gratefully acknowledged. I also wish to thank Anita Lustenberger for her help in developing the questionnaire.

I am very grateful to the following individuals and their laboratories for providing data for the amniocentesis survey: E. Allen, M. Alonzo, D. Arakaki, D. Arthur, P. Bader, D. Borgaonkar, R. Breg, J. Brown, H. Chen, M. Cohen, B. Crandall, C. Disteche, R. Falk, E. Gendel, M. Golbus, F. Grass, R. Greenstein, T. Hadro, J. Higgins, J. Jackson, M. Jenkins, O. Jones, S. Kaffe, B. Kaiser-McCaw, N. Kardon, P. Katayama, C. King, P. Kohn, F. Kappitch, A. Kutasovic, J. Lanman, E. Lieber, G. Livingston, E. Magenis, P. Martens, M. Mennute,

P. Minka, T. Mohandas, C. Moore, G. Pai, I. Paika, C. Palmer, S. Patil, J. Priest, H. Punnet, Q. Qazi, K. Richking, S. Riley, A. Robinson, I. Rosenthal, H. Rothschild, M. Sandstrom, W. Sanger, B. Say, A. Schonhaut, L. Shapiro, S. Sherman, J. Simpson, S. Soukup, G. Stetten, K. Taysi, M. Temple, A. Tharapel, C. Trunca, J. Tucker, L. Veeck, J. Waurin, B. Weisskopf, W. Wertelecki, A. Willey, S. Wolman, D. Wurster-Hill, H. Wyandt, D. Van Dyke, S. Young.

REFERENCES

1. Friedrich V, Nielsen J. Bisatellited extra small metacentric chromosome in newborns. *Clin Genet* 1974; 5: 23-31.

2. Walzer S, Breau G, Gerald P.S. A chromosome survey of 2,400 normal newborn infants. *J Pediat* 1969; 74: 438-48.

3. Weber FM, Dooley RR, Sparkes RS. Anal atresia, eye anomalies and an additional small abnormal metacentric chromosome. (47, XX, Mar+): Report of a case. *J Pediat* 1970; 76: 594-97.

4. Wisniewski L, Hassold J, Heffelfinger J, Higgins JV. Cytogenetic and clinical studies in five cases of inverted duplication (15). *Hum Genet* 1979; 50: 259-70.

5. Van Dyke DL, Weiss L, Logan M, Pai GS. The origin and behavior of two isodicentric bisatellited chromosomes. *Am J Hum Genet* 1977; 29: 294-300.

6. Breg WR, Miller DA, Allderdice PW, Miller OJ. Identification of translocation chromosomes by quinacrine fluorescence. *Am J Dis Child* 1972; 123: 561-64.

7. Jacobs PA. Correlation between euploid structural rearrangements and mental subnormality in humans. *Nature* 1974; 249: 164-65.

8. Jacobs PA, Matsuura JS, Meyer M, Newlands IM. A cytogenetic survey of an institution for the mentally retarded. 1. Chromosome abnormalities. *Clin Genet* 1978; 13: 37-60.

9. Jacobs PA. Mutation rates for structural chromosome rearrangements in man. *Amer J Hum Genet* 1981; 33: 44-54.

10. Funderburk SJ, Spence MA, Sparkes RS. Mental retardation associated with "balanced" chromosome rearrangements. *Amer J Hum Genet* 1977; 29: 136-41.

11. Jacobs PA, Frackiewicz A, Law P. Incidence and mutation rates of structural rearrangements of the autosomes in man. *Ann Hum Genet* 1972; 35: 301-19.

12. Patil SR, Lubs HA, Kinberling WJ, *et al.* Chromosomal abnormalities ascertained in a collaborative survey of 4,342 seven and eight year old children: frequency, phenotype and epidemiology. *In:* Hook EB, Porter IH, eds. Population cytogenetics. New York: Academic Press, 1977: 103-31.

13. Breg WR. Euploid structural rearrangements in the mentally retarded. See Ref. 12, pp.99-102.

14. Broggar A. Concerning translocation as a cause of mental retardation. *In:* International Copenhagen Congress Study of Mental Retardation. 1964: 128-131.

15. Carrel RE. Chromosome survey of moderately to profoundly retarded patients. *Am J Ment Def* 1973; 77: 616-22.

16. Cassiman JJ. Sex chromatin and cytogenetic survey of 10,417 adult males and 357 children institutionalized in Belgian institutions for mentally retarded patients. *J Humangenetic* 1975; 28: 43-8.

17. Chen ATL, Sergovitch FR, McKim JS, Barr ML, Gruber D. Chromosome studies in full-term low-birth-weight, mentally retarded patients. *J Ped* 1970; 76: 393-98.

18. Coco R, Penchaszadeh VB. Frequency of chromosomal aberrations in 131 patients with multiple congenital malformations and mental retardation. *J Ped* 1976; 89: 325.

19. Corey MJ. Structural aberrations of autosomes in a mentally retarded population. *Am J Ment Def* 1971; 75: 487-98.

20. Daly, RF. Chromosome aberrations in 50 patients with iodiopathic mental retardation and in 50 control subjects. *J Ped* 1970; 77: 444-53.

21. Doyle CT. The cytogenetics of 90 patients with iodiopathic mental retardation/malformation syndromes and of 90 normal subjects. *Hum Genet* 1976; 33: 131-416.

22. Erdtmann B, Salvano FM, Mattevi MS. Chromosome studies in patients with congenital malformations and mental retardation. *Humangenetik* 1975; 26: 297-306.

23. Faed M. A chromosome survey of a hospital for the mentally subnormal. *J Med Genet* 1972; 9: 470-72.

24. Fujita H, Fujita K. A cytogenetic survey on mentally retarded children. *Jap J Hum Genet* 1974; 19: 175-76.

25. Iivanainen M, Gripenberg U. Clinico-neurological findings in connection of three chromosomal aberrations: An extra chromosome in group E, a D/C translocation and an unusually long B group chromosome. *Acta Neurol Scand (Suppl)* 1967; 43(31): 53-4.

26. Jacobs PA. Correlation between euploid structural chromosome rearrangements and mental subnormality in humans. *Nature* 1974; 249: 164-65.

27. Lubs HA, Lubs ML. New cytogenetic techniques applied to a series of children with mental retardation. *In:* Nobel Symposia XXIII, 1973: 241-50.

28. Magnelli NC, Cytogenetics of 50 patients with mental retardation and multiple congenital anomalies and 50 normal subjects. *Clin Genet* 1976; 9: 169-82.

29. Newton MS, Cunningham C, Jacobs PA, Price WH, Fraser IA. Chromosome survey of a hospital for the mentally retarded. Part 2: Autosome abnormalities. *Clin Genet* 1972; 3: 226-48.

30. Speed RM, Johnston AW, Evans HJ. Chromosome survey of total population of mentally subnormal in Northeast of Scotland. *J Med Genet* 1976; 13: 295-306.
31. Sutherland GR, Weiner S. Chromosome studies in a mental deficiency hospital: Total ascertainment. *Aust J Ment Ret* 1971; 1: 246-47.
32. Sutherland GR, Murch AR, Gardiner AJ, Carter RF, Wiseman C. Cytogenetic survey of a hospital for the mentally retarded. *Hum Genet* 1976; 34: 231-45.
33. Thornburn MJ, Martin PA. Chromosome studies in 101 mentally handicapped Jamaican children. *J Med Genet* 1971; 8: 59-64.
34. Yanagisawa S. Cytogenetic studies on the mentally retarded children. *Acta Paediatr, Jpn* 1968; 10: 30.

CHROMOSOME MOSAICISM AND PSEUDOMOSAICISM IN PRENATAL CYTOGENETIC DIAGNOSIS

Lillian Y.F. Hsu
Theresa E. Perlis

INTRODUCTION

One of the most difficult problems in prenatal cytogenetic diagnosis is the differentiation between true chromosome mosaicism and pseudomosaicism. This presentation will primarily review the data acquired from the recent United States of America (US) national survey on chromosome mosaicism and pseudomosaicism with emphasis in the following three major areas:

1. The incidence of true chromosome mosaicism and pseudomosaicism in prenatal diagnosis.
2. The frequency of involvement of specific chromosomes or chromosome complements in pseudomosaicism including both numerical and structural aberrations.
3. Information on the phenotypic manifestation and cytogenetic confirmation studies in cases with diagnosis of true mosaicism.

US SURVEY

During the 1979 international workshop on prenatal diagnosis[1] which discussed the past, present and future of prenatal diagnosis, one of us (Dr. L. Y.F. Hsu) was requested to assume the responsibility of undertaking a survey in the US on chromosome mosaicism and pseudomosaicism in prenatal cytogenetic diagnosis. The project was initiated in 1980 when approximately 150 cytogenetic laboratories in the US were invited to participate. Seventy-five responded positively. By June 1981, 59 laboratories (see Participating Laboratories listed at the end of this chapter) submitted their data on mosaicism and/or pseudomosaicism from approximately 60,000 genetic amniocenteses.

In the survey, a diagnosis of true chromosome mosaicism was made when an identical chromosome abnormality was detected in multiple independently cultured vessels or at the very least two cultured vessels. However, when the aberrant karyotype was restricted to one culture flask, pseudomosaicism was

Editor's note: Tables appear at the end of this chapter.

diagnosed. Cases classified as pseudomosaicism were subdivided into "single cell" or "multiple cell" categories according to the number of cells involved in an identical abnormality. For those laboratories which used *in situ* culture and harvesting method, three types of pseudomosaicism were described:[2] (a) one cell or one region of a clone showing an aberrant karyotype, (b) a single colony in which all cells have the same abnormality, and (c) several colonies from one culture vessel showing the identical aberrant karyotype. In this survey, these three types were grouped into two major categories: (a) single cell or single clone, and (b) multiple cells or multiple clones.

In our survey, 42 laboratories utilized a closed system, *i.e.* flask culture and trypsinized cells for harvests. Thirteen laboratories used an open system, *i.e.* petri dishes and *in situ* harvesting method. Four laboratories used a combination of both methods. The vast majority (48 laboratories) routinely harvested two or more culture vessels. Twenty-nine laboratories analysed ten cells per culture vessel. Twenty-eight laboratories routinely examined a total of 20 cells. G- and/or Q-banding was used routinely in 57 laboratories while two laboratories preferred R-banding primarily.

I Incidence of True Mosaicism

The frequency of true mosaicism derived from the data of 59 US cytogenetic laboratories which participated in this survey and from data of two other investigators[3] (Hsu, LYF. Unpublished data from the Cytogenetic Laboratory of Mount Sinai School of Medicine) previously collected, is shown in Table 1. The frequencies range from 0–0.89%. Zero percentage was reported by 17 laboratories of which 11 had a study population of less than 500 cases each. Thirty-seven laboratories reported a frequency between 0.1–0.5%; 28 of these studied more than 500 cases (Table 2). Apparently there is no correlation between the frequency of mosaicism and the size of the study population. In a total of 62,279 cases of genetic amniocenteses, there were 156 cases of true chromosome mosaicism representing an overall incidence of 0.25%, or 2.5 cases per every 1,000 studied (Table 3). The frequency of mosaicism was not significantly different between laboratories that used the closed flask system and laboratories that used petri dishes and *in situ* method (Flask 104/40, 219 = 0.258%; Dish 29/16,600 = 0.235%). The frequency of chromosome mosaicism diagnosed prenatally was much higher than that found in the large consecutive live birth studies which showed an incidence of less than one per 1,000 (18/42, 989 = 0.04%).[4-7] Conceivably, the lower frequency of mosaicism found in the live birth studies could be partially attributed to the methodology bias, since only a small number of cells were analysed and chromosome mosaicism was of no major concern or interest.

II Frequency of Pseudomosaicism

The frequencies of pseudomosaicism from all participating laboratories are tabulated in Table 4, including occurrence of (a) all types of pseudomosaicism,

(b) multiple cells only, and (c) single cells only. The study population varies from 84 to 4,000, and the frequencies of pseudomosaicism (of all types) range from 0–15.52%. The frequencies of pseudomosaicism apparently are not related to the number of patients included in each study. The majority of laboratories reported a study population of less than 1,000 and a frequency of 0.1–7.0% (Table 5). Eleven laboratories reported a frequency of 0.1–1.0%; eight laboratories 1.1–2.0%; six laboratories 2.1–3.0%; and five laboratories 3.1–4.0%.

For pseudomosaicism involving multiple cells or clones, 32 laboratories reported a frequency of 0.1–1.0%, and 11 laboratories reported 1.1–3.05% (Table 6).

The overall frequency of all types of pseudomosaicism is 3.24% in a total of 33,255 genetic amniocenteses studied (Table 3). The overall frequency of pseudomosaicism involving multiple cells or multiple clones is 0.7% in a total of 48,442 cases studies (Table 3).

III Frequency of Chromosome Involvement in Pseudomosaicism

A. PSEUDOMOSAICISM WITH NUMERICAL ABNORMALITIES

Trisomy: It is known that one of the most frequent *in vitro* findings in cultured human amniotic fluid cells is trisomy 2.[3,8,9] The US survey has further confirmed this finding. In a total of 444 cases with pseudomosaicism involving a single cell (or clone), or multiple cells (or clones) with trisomy, chromosome 2 was involved 22.7% of the time. The other three frequently involved chromosomes were 20, 7 and X (Table 7). The frequent occurrence of trisomy 2 becomes even more striking when the frequencies of trisomic chromosomes were analysed in 90 cases with pseudomosaicism involving multiple cells, while the five other frequently involved chromosomes were 7, 9, 17, 20 and X (Table 8).

Monosomy: Since finding a single cell with 45 chromosomes with one chromosome missing may simply represent an artifact or a random loss of a chromosome during *in vitro* processing, only findings of the same type of monosomy involving multiple cells are useful in assessment of the frequencies of monosomic chromosomes in pseudomosaicism. In a total of 78 cases missing the same chromosome in multiple cells, the six most frequently involved chromosomes were the X, Y, 21, 22, 17 and 19 (Table 9).

B. PSEUDOMOSAICISM WITH STRUCTURAL ABNORMALITIES

From a total of 393 instances of pseudomosaicism involving structural aberrations of a specified short arm (p), or long arm (q), this survey showed a very high correlation (.94) between relative chromosome size and frequency of involvement in structural rearrangement. Using the chi-square test for goodness of fit with a five percent critical level, it was found that the frequency of involvement of chromosomes in short arm aberrations could be attributed to relative size alone. However, for long arm aberrations, the frequency of chromosome 1 involvement was far higher than expected, although for the remaining chromosomes the frequency of long arm involvement could again be attributed to relative size alone. It should be noted that the acquired data does not identify the specific regions or bands at which the breakpoints occurred. Since precise identification of the breakpoints involved in rearrangement was either not carried out or the information not provided in this survey, the available data cannot identify any specific chromosome region or band.

IV Follow-up Data from Cases with Pseudomosaicism

Of all the cases with pseudomosaicism involving either single cell or multiple cells with aberrant karyotypes, only 754 cases had available records for the outcome of the pregnancy. There was one case in which one culture flask had 44 cells with a 45, X karyotype and six cells with 46, XX, while all 55 cells from the other flask showed a 46, XX constitution. Although this pregnancy resulted in a phenotypically normal female infant, two out of 50 lymphocytes from blood culture showed 45, X. This infant is most likely a true mosaic for 45, X cell line. In one other case, one out of five flasks showed 7/21 cells with trisomy 20. The patient elected termination and trisomy 20 was found in 3/85 cells (from one flask) in the abortion fluid from an otherwise normal appearing male fetus.

Eleven other cases resulted in live births with a variety of defects (Table 10) apparently not related to chromosome abnormalities. The frequency for major anomalies (11/754 or 1.5%) is well within the general expectation in live births. Thirteen other pregnancies with single cell pseudomosaicism resulted in stillbirths. Again, this frequency does not exceed the expected range of stillbirths in pregnancies of the advanced maternal age group.[10]

V Phenotypic Manifestations and Cytogenetic Confirmation of Cases with True Mosaicism

The following information includes chromosome mosaicism diagnosed prenatally from this survey as well as from other documented cases.

A. SEX CHROMOSOMES

The incidence of sex chromosome mosaicism in prenatal diagnosis is almost equal to that for all of the autosomes. A total of 73 cases of sex chromosome mosaicism has been collected including 21 cases of 45, X/46, XX; 17 cases of 45, X/46, XY; 13 cases of 46, XY/47, XXY; seven cases of 45, X/47, XXX; six cases of 46, XX/47, XXX; three cases of 45, X/46, Xr (X); two cases of 45, X/46, Xi (Xq); one case each for 45, X/46, XX/46, Xi (Xq); 46, XY/47, XYY; 46, XY/47, XYY/48, XYYY; and 46, XY/46, XYq− (Table 11).

Of 21 cases of 45, X/46, XX mosaicism, nine were terminated, 15 had information on the phenotypes, 11 were reported to be phenotypically normal females and four were described as abnormal. One live birth had noticeable Turner's stigmata and cytogenetic follow-up studies showed only 45, XO cells. Although the majority of cases with XO/XX mosaicism appeared to be phenotypically normal either at birth or after termination, one must realize that even in patients with 45, X without mosaicism, the major features of Turner syndrome (such as short stature and sexual infantilism) are not manifested either before or at birth. Furthermore cases with 45, X/46, XX mosaicism diagnosed prenatally present an unbiased group of patients, since most postnatally diagnosed cases were ascertained through their clinical manifestations (Table 11).

There were 13 other X chromosome mosaics including a 45, X cell line. From these a total of seven cases had known outcomes (Table 11) and only one was described as phenotypically abnormal.

Of the 17 cases with 45, X/46, XY mosaicism, ten pregnancies were terminated; except for one abortus reported to be a phenotypically normal female, the nine other abortuses and seven live births were reported to be normal males. In no case was ambiguous genitalia noted. In 13 successful cytogenetic follow-up studies, mosaicism was confirmed in eight cases, whereas normal chromosome complements of 46, XY were found in five other cases (Table 11).

Of 13 cases with 46, XY/47, XXY mosaicism, six were terminated. In ten of these cases the outcome is known and all ten were grossly normal male abortuses or live births. Only long term follow-up of the six live births will provide us with information regarding any clinical manifestations of Klinefelter's syndrome (Table 11).

B. AUTOSOMES

Exclusive of 23 cases of small marker chromosome mosaicism, there were 89 cases of autosomal mosaicism, including 21 cases of 46/47, +21; 20 cases of 46/47, +20; six cases of 46/47, +9; five cases of 46/ 47, +8; four cases of 46/47, +18; two cases of 46/47, +13; two cases of diploid/ triploid mosaicism; 17 cases of numerical abnormalities involving various autosomes and 12 cases of mosaicism involving a structural rearrangement (Table 12).

Of 21 cases with trisomy 21 mosaicism, 12 were terminated. Thirteen cases had available information regarding the phenotypic manifestation; nine of those were described as abnormal. Cytogenetic confirmation was achieved in 13 cases out of 15 successfully cultured, whereas only normal chromosomes were demonstrated in the other two cases (Table 12).

A total of 20 cases had trisomy 20 mosaicism. Twelve were terminated. Seventeen resulted in either phenotypically normal abortuses or live births. Three abortuses were described as abnormal. In these fetuses, trisomy 20 was confirmed in fetal tissues (kidney cells in two and rectal cells in one). Of 16 successful follow-up cytogenetic studies, mosaicism was confirmed in six cases. Thus far this type of mosaicism remains a problem in genetic counseling (Table 12).

All six cases diagnosed with trisomy 9 mosaicism elected to terminate the pregnancies. Two abortuses were phenotypically abnormal; cytogenetic confirmation was successful in one of these. Four resulted in apparently normal fetuses; trisomy 9 mosaicism was confirmed in one of these, and cytogenetic analysis of the three other fetuses showed normal karyotypes only (Table 12).

All five cases with trisomy 8 mosaicism terminated their pregnancies. Postmortem examination showed three apparently normal abortuses, one abnormal fetus, and in one case the phenotype was unknown. Cytogenetic confirmation was achieved in two cases; two showed only normal chromosomes in the fetal tissues, and one abortus was not studied (Table 12).

Of four cases with trisomy 18 mosaicism, all were terminated. Three were grossly abnormal and no information was available on the fourth. Cytogenetic confirmation was successful in all four cases. Both cases with trisomy 13 mosaicism were terminated and resulted in phenotypically abnormal abortuses. Cytogenetic confirmation was possible in only one (Table 12). In the two cases with diploid/triploid mosaicism, one was terminated and one resulted in a stillbirth. Both fetuses were obviously abnormal and cytogenetic confirmation was achieved in both (Table 12).

There were six cases of questionable monosomy mosaicism including one case each for monosomy 9, 17, 19, and 20, and two cases of monosomy 22. The diagnoses of those six cases were based on the finding of one monosomic cell in each of two culture flasks. No case was terminated. The outcome of those pregnancies showed two apparently normal live births, three lost to follow-up, and one case with 46/45, −22 resulted in a neonatal death with multiple abnormalities. The mosaicism of monosomy 22 was confirmed from the blood culture of the neonate. Therefore, monosomy should not be overlooked if detected from two or more culture vessels (Table 12).

Of 15 cases with mosaicism involving an autosomal structural aberration, nine included at least one cell line with unbalanced chromosome complement. Of 14 with available follow-up information, three abortuses and one live birth were phenotypically abnormal (all contained at least one cell line with unbalanced chromosome complement) and ten were grossly normal. Cytogenetic confirmation was achieved in only two of the four phenotypically abnormal cases. Of the remaining 13 cases, normal chromosomes were found in nine, and cytogenetic studies were either not successful or not carried out in four other cases (Table 12).

C. MARKER CHROMOSOME

Twenty-three cases were mosaic for a small marker chromosome. Twelve were terminated. Of 17 cases with available information regarding phenotypic manifestations, three were described as abnormal, 16 had successful follow-up cytogenetic studies, mosaicism was confirmed in 13 cases (Table 13). In three cases, mosaicism for an identical marker chromosome was demonstrated in one phenotypically normal parent (laboratories *c, i, q*). Therefore, cytogenetic studies should be carried out in the parents so as to provide additional information for genetic counseling.

D. SUMMARY OF PRENATAL DIAGNOSIS OF CHROMOSOME MOSAICISM

Of 185 cases of chromosome mosaicism, 89 cases (48.1%) were autosomal, 73 cases (39.5%) involved a sex chromosome and 23 cases (12.4%) were mosaic for a small marker chromosome of unidentified origin (Table 14). As expected the frequency of noticeable phenotypic abnormalities was highest (37.8%) in the autosomal mosaics and lowest (10.5%) in the sex chromosome mosaics. The average rate for cytogenetic confirmation was 70%.

VI Number of Culture Vessels Used and Cells Analysed to Distinguish Pseudomosaicism from True Mosaicism

In a total 754 cases of pseudomosaicism with known pregnancy outcome, the average number of culture vessels studied and the average number of cells examined are slightly different for the different types of pseudomosaicism. In pseudomosaicism involving multiple cells, more laboratories studied three or more culture vessels in comparison to the frequencies of flasks studied for single cell pseudomosaicism, also more cells were analysed in cases of multiple cell pseudomosaicism. A much larger number of laboratories analysed 21 to 40 cells or more from both the flasks with normal cells and abnormal cells than for single cell pseudomosaicism.

VII Routine Practice in Most New York State Laboratories for the Differentiation of Mosaicism from Pseudomosaicism

Routinely three or more culture flasks are established for culture and two culture vessels are harvested for analysis. Ten cells are examined from each flask. If all twenty cells are normal, a report of normal result is sent out. In cases with numerical abnormality, a third flask is harvested and 20 cells from each of the three flasks are analyzed to search for the identical aberrant karyotype. If the same abnormality is detected in two or more flasks, a diagnosis of true chromosome mosaicism is then made. In cases with structural aberrations, ten additional cells from the normal flask will be analysed to search for an identical aberration. If no other cell shows the same abnormality, a diagnosis of pseudomosaicism is entertained. It must be emphasized that this is the minimum requirement for differentiation between pseudomosaicism and true mosaicism.

PRELIMINARY REPORT OF CANADIAN AND EUROPEAN SURVEYS

Worton (personal communication) collected data from 12,680 genetic amniocenteses from 14 different cytogenetic laboratories in Canada. These cases were studied between 1971 and 1980. In the Canadian series, the frequency of chromosome mosaicism was 0.41% (52/12, 680). This frequency was higher than the US survey, possibly due to the fact that the Canadian frequency also included cases with an identical abnormality detected from multiple colonies but restricted to one culture vessel. Those cases would have been classified as pseudomosaicism in the US survey.

The incidence of single cell pseudomosaicism in the Canadian survey was 10.236%(1,298/12,680), and the incidence of multiple cell pseudomosaicism restricted to one colony or one flask was 2.610%(331/12,680).

In approximately 200 cases with single cell pseudomosaicism involving trisomy where the extra chromosome was identified, chromosome 2 was again noted to be the most frequent one. Chromosomes 7 and 21 were the next most frequently involved chromosomes in that survey.

In a total of 258 cases of multiple cell pseudomosaicism, chromosome 2 was the most frequently involved one in trisomy and chromosomes 21, 22, 17, 19, and Y were most frequently involved in monosomy.

In cases of both single cell (522 cases) and multiple cells (96 cases) with structural abnormalities found within one flask or one colony, the frequencies of chromosome involvement did not correlate with the size of the chromosome. There was no correlation between the data collected from the single cell and from the multiple cells.

In the European study, Lindsten (personal communication) obtained data from 36,659 genetic amniocenteses from 32 cytogenetic laboratories. The frequency of chromosome mosaicism was 0.095% which was lower than the US survey. The incidence of single cell pseudomosaicism was 2.93% and multiple cell pseudomosaicism was 0.59% The European survey also indicated that there was no difference in the incidence of mosaicism or pseudomosaicism when different culture and harvest methods were utilized. Trisomy 2 was once again shown to be the most frequent finding in pseudomosaicism.

CONCLUDING REMARKS

A last question remains: Is there a better method for the differential diagnosis of chromosome mosaicism from pseudomosaicism? Since cultured amniotic fluid cells are grown out clonally from a few original fetal cells, the total number of cells examined from a pooled trypsinized flask does not represent the actual number of the original cells. Theoretically, analyzing cells from colonies with the *in situ* harvesting method is more effective in determining the frequency of abnormal clones rather than the number of abnormal cells in diagnosing mosaicism. However, experience showed multiple affected clones in *one* culture vessel cannot equate or replace finding of the same abnormality in at least *two* culture vessels.[1,2,8] The open culture system and *in situ* harvest method present an increased risk of contamination and has thus become a less practical and rather unpopular culture method. Even the laboratories which employ *in situ* method seldom examine more than 30 colonies.

According to statistics,[26] in order to detect five percent mosaicism, one should examine 60 colonies. The routine analysis of 20 cells from two flasks is a practical method but far from ideal, because 20 analyzed cells represents much less than 20 colonies and can only achieve an effect of detecting mosaicism somewhere from 14% to 50% at the 95% confidence level. Therefore, until a better and more practical method such as a closed system with *in situ* harvest is established and improved, one will not be better equipped to deal with such a problem. Even then this problem of mosaicism will not disappear completely, because as we all know in biology nothing is 100%.

ADDENDUM

This presentation was not designed to survey the frequencies of maternal cell contamination. From the preliminary data of this laboratory, ten instances (Benn and Hsu, unpublished data) of maternal cell contamination were observed in the first 2,000 cases studied (representing 0.5%). In almost all cases it was *not* associated with grossly bloody taps. In three cases the maternal cells were found in multiple flasks. The chromosome polymorphism comparisons made of mother's cells were helpful in the majority of cases.

ACKNOWLEDGMENTS

This study was supported by the New York State Department of Health and by the New York City Department of Health.

The authors are grateful to all 59 participating cytogenetic laboratories throughout the country for submitting their valuable data. Special thanks to Peter A. Benn, Ph.D., for his input in designing the mosaicism forms and valuable criticism in the content; Hody Tannenbaum for editing, Amy G. Schonnaut and Beth Siegel for recording the Prenatal Diagnosis Laboratory data and Holger Hansen, M.D., for his valuable consultation in statistics, and Carmen Vazquez for preparation of the manuscript.

PARTICIPATING LABORATORIES

a   Allen E. University of Vermont, Burlington, Vermont
b   Arthur DC, Millard C, Pierpont ME. University of Minnesota Hospitals,
     Minneapolis, Minnesota
c   Bannerman R, Fadness P. Children's Hospital, Buffalo, New York
d   Brown JA. Medical College of Virginia, Richmond, Virginia
e   Byrd, JR. Medical College of Georgia, Augusta, Georgia
f   Chen H, Yu CW. Louisiana State University Medical Center, Shreveport,
     Louisiana
g   Cohen M. Children's Memorial Hospital, Chicago, Illinois
h   Comings DL, Teplitz R, Richkind K. City of Hope National Medical
     Center, Duarte, California
i   Crandall B. University of California at Los Angeles, Los Angeles,
     California
j   Dewald G. Mayo Clinic, Rochester, Minnesota
k   Dumars K. University of California Irvine Medical Center, Orange,
     California
l   Evans TN. C.S. Mott Center Wayne State University School of Medicine,
     Detroit, Michigan
m   Golbus M. University of California Medical Center, San Francisco,
     California
ǹ   Grass F. Charlotte Memorial Hospital and Medical Center, Charlotte,
     North Carolina
o   Greenstein R, Gasparini R. University of Connecticut School of Medicine,
     Farmington, Connecticut
p   Hecht F, Kaiser-McCann B. Genetics Center of Southwest Biomedical
     Research Institute, Tempe, Arizona
q   Hsu LYF, Benn PA. Prenatal Diagnosis Laboratory of New York City,
     Medical and Health Research Association of New York City.
r   Ing P. University of Miami, Mailman Center for Child Development,
     Miami, Florida
s   Jackson J. University of Mississippi Medical Center, Jackson, Mississippi
t   Jones OW. University of California-San Diego, La Jolla, California
u   Kaback MM, Mohandas T. Harbor-U.C.L.A. Medical Center, Torrance,
     California
v   Kaffe S, Desnick R. Mount Sinai School of Medicine, New York,
     New York
w   Kardon N. North Shore University Hospital, Manhasset, New York
x   Kohn P. J. Hillis Miller Health Center, Gainesville, Florida
y   Kousseff B, Hadro TA. Southern Illinois University School of Medicine,
     Springfield, Illinois
z   Latt S, Sandstrom M McH. Brigham Women's Hospital, Boston,
     Massachusetts
aa   Lieber E. Long Island Jewish-Hillside Medical Center, New Hyde Park,
     New York
bb   Lozzio, CB. University of Tennessee, Birth Defects Evaluation Center,
     Knoxville, Tennessee
cc   Martin G, Norwood T, Disteche C, Salk D. University of Washington,
     Seattle, Washington

dd Meisner, LF. University of Wisconsin, Madison, Wisconsin
ee Mennuti MT. Hospital of the University of Pennsylvania, Philadelphia, Pennsylvania
ff Miller WA. Massachusetts General Hospital, Boston, Massachusetts
gg Milunsky A. Eunice Kennedy Shriver Center, Waltham, Massachusetts
hh Ming PM. Temple University School of Medicine, Philadelphia, Pennsylvania
ii Minka DF, Antley R. Methodist Hospital, Indianapolis, Indiana
jj Moore, CM. University of Texas, Medical School, Houston, Texas
kk Patil SR. University of Iowa Hospital, Iowa City, Iowa
ll Priest JH. Emory University, Atlanta, Georgia
mm Punnett H. St. Christopher's Hospital for Children, Philadelphia, Pennsylvania
nn Riley S, Saul R. Greenwood Genetic Center, Greenwood, South Carolina
oo Robinson A, Ball S. National Jewish Hospital and Research Center, Denver, Colorado
pp Rosenthal I. University of Illinois at the Medical Center of Chicago, Chicago, Illinois
qq Sanger WG. University of Nebraska Medical Center, Omaha, Nebraska
rr Say B. Children's Medical Center, Tulsa, Oklahoma
ss Schmickel R. University of Michigan Medical School, Ann Arbor, Michigan
tt Sciorra L. College of Medicine and Dentistry of New Jersey, Rutgers Medical School, Piscataway, New Jersey
uu Shapiro L. Letchworth Village Developmental Center, Thiells, New York
vv Simpson JL, Martin A. Prentice Women's Hospital, Northwestern University, Chicago, Illinois
ww Soukup S. Children's Hospital Research Foundation, Cincinnati, Ohio
xx Stetten G. John Hopkins University School of Medicine, Baltimore, Maryland
yy Trunca C. State University of New York at Stony Brook, New York
zz Van Dyke DL. Henry Ford Hospital, Detroit, Michigan
aaa Vinson PC. University of Alabama at Birmingham, Birmingham, Alabama
bbb Warburton D. Columbia University College of Physicians and Surgeons, New York, New York
ccc Wertelecki W. University of South Alabama Medical School, Mobile, Alabama
ddd Willey A. New York State Department of Health, Birth Defects Institute, Albany, New York
eee Wolman SR. New York University Medical Center, New York, New York
fff Wyandt, HE. University of Virginia Medical Center, Charlottesville, Virginia
ggg Ying KL. Valley Children's Hospital, Fresno, California

Table 1
Frequency of Chromosome Mosaicism in Prenatal Cytogenetic Diagnoses

Lab.*	Total Cases Studied	No. of Cases	%	Lab.*	Total Cases Studied	No. of Cases	%
a	167	0		ee	1104	2	0.18
b	944	0		ff	2455	4	0.16
c	474	2	0.42	gg	3440	10	0.29
d	791	1	0.13	hh	133	0	
e	725	1	0.14	ii	237	1	0.42
f	84	0		jj	488	0	
g	2600	1	0.04	kk	950	1	0.11
h	357	1	0.28	Ref. 3	1100	2	0.18
i	4000	7	0.18	ll	968	4	0.41
j	116	0		mm	308	2	0.65
k	522	3	0.57	nn	309	2	0.65
l	684	2	0.29	oo	1000	2	0.20
m	5000	22	0.44	pp	160	1	0.63
n	265	1	0.38	qq	527	1	0.19
o	177	0		rr	126	0	
p	275	0		ss	456	0	0.22
Hsu**	1401	5	0.36	tt	934	3	0.32
q	2088	9	0.43	uu	1482	5	0.34
r	1065	0		vv	887	0	
s	422	1	0.24	ww	700	3	0.43
t	3376	9	0.27	xx	719	0	
u	867	1	0.12	yy	100	0	
v	768	2	0.26	zz	1305	3	0.23
w	2027	5	0.25	aaa	2360	6	0.25
x	101	0		bbb	1890	9	0.48
y	225	2	0.89	ccc	300	0	
z	632	0		ddd	1300	4	0.31
aa	382	1	0.26	eee	239	1	0.42
bb	405	1	0.25	fff	507	0	
cc	2264	10	0.44	ggg	510	1	0.20
dd	2081	1	0.05				

*See Participating Laboratories list.
**Hsu, LYF, unpublished data, Mount Sinai.

TABLE 2

Number of Laboratories and Frequencies of Chromosome Mosaicism

Frequency of Mosaicism %	≤500 Cases	501 to 1000	1001 to 1500	1501 to 2000	2001 to 2500	2501 to 3000	3001 to 4000	4001 to 5000	Sub-Total
0	11	5	1						17
0.01–0.10					1	1			2
0.11–0.20		7	2		1		1		11
0.21–0.30	5	2	1		2		2		12
0.31–0.40	1	1	3						5
0.41–0.50	3	2		1	2			1	9
0.51–0.60		1							1
0.61–0.70	3								3
0.71–0.80									
0.81–0.90	1								1
Subtotal	24	18	7	1	6	1	3	1	61

TABLE 3
Frequency of Mosaicism and Pseudomosaicism Among All Cases Studied

	Mosaicism	Pseudo. all types	Pseudo. mult. cells	Pseudo single cell
No. cases	62,279	33,255	48,442	30,754
No. cases mos./pseudo.	156	1,080	340	758
Ave. freq. mos./pseudo.	0.250%	3.248%	0.702%	2.465%
Range non-zero freqs. (all labs)	0.04%– 0.89%	0.19%– 15.52%	0.12%– 11.21%	0.09%– 11.49%

TABLE 4
Frequency of Pseudomosaicism

		No. of Occurrences					
	Total	All types		Multiple cells		Single cell	
Lab.*	Cases	No.	%	No.	%	No.	%
a	167	3	1.80	1	0.60	2	1.20
b	944	49	5.19	21	2.22	28	2.97
c	474	24	5.06	20	4.22	4	0.84
d	791		NR**	16	2.02		NR
e	725	21	2.90	6	0.83	15	2.07
f	84	1	1.19	1	1.19	0	
g	2600		NR	6	0.23		NR
h	357	25	7.00	9	2.52	16	4.48
i	4000		NR	13	0.33		NR
j	116	18	15.52	13	11.21	5	4.31
k	522	6	1.15	1	0.19	5	0.96
l	778	6	0.77	4	0.51	2	0.26
n	265	5	1.89	4	1.51	1	0.38
o	177	5	2.82	3	1.69	2	1.13
p	275	10	3.64	0		10	3.64
Hsu†	1401	19	1.35		NS††		NS
q	1632	96	5.88	16	0.98	80	4.90
r	1065	2	0.19	0		2	0.19
s	422		NR	2	0.47		NR
t	3376		NR	4	0.12		NR
u	867	17	1.96	4	0.46	13	1.50
v	851		NR	2	0.24		NR
w	888	110	12.39	8	0.90	102	11.49
x	101	4	3.96	0		4	3.96

TABLE 4, continued

| | | No. of Occurrences | | | | | |
| | Total | All types | | Multiple cells | | Single cell | |
Lab.*	Cases	No.	%	No.	%	No.	%
y	225	15	6.67	0		15	6.67
z	632	75	11.87	19	3.01	56	8.86
aa	382	3	0.79	2	0.52	1	0.26
bb	405	13	3.21	5	1.23	8	1.98
cc	2264	45	1.99	4	0.18	41	1.81
dd	2081		NR	10	0.48		NR
ee	1104	3	0.27	2	0.18	1	0.09
gg	3440	15	0.44	5	0.15	10	0.29
hh	133	1	0.75	0		1	0.75
ii	237	1	0.42	0		1	0.42
jj	488		NR	3	0.61		NR
kk	950	20	2.11	4	0.42	16	1.68
Ref. 3	1100	29	2.63		NS		NS
ll	968	117	12.09	21	2.17	96	9.92
mm	333	4	1.20	4	1.20	0	
oo	1000	57	5.70	14	1.40	43	4.30
pp	160	0					
rr	126	0					
ss	456	15	3.29	4	0.88	11	2.41
tt	934	61	6.53	8	0.86	53	5.67
uu	1482	5	0.34	4	0.27	1	0.07
vv	887	29	3.27	2	0.23	27	3.04
ww	700	0					
xx	719		NR	6	0.83		NR
yy	100	1	1.00	1	1.00	0	
zz	1305	13	1.00	6	0.46	7	0.54
aaa	2360		NR	4	0.17		NR
bbb	458	11	2.40	3	0.66	8	1.75
ccc	300	7	2.33	3	1.00	4	1.33
ddd	1080	53	4.91	44	4.07	9	0.83
eee	239	19	7.95	2	0.84	17	7.11
fff	507	4	0.79	1	0.20	3	0.59
ggg	510	43	8.43	5	0.98	38	7.45

*See Participating Laboratories list.
**NR = Not reported.
†Hsu, LYF, previously collected data, Mount Sinai.
††NS = Not separated.

TABLE 5
Number of Laboratories and Frequencies of Pseudomosaicism
(Single and Multiple Cells) Detected

Frequency of Pseudomo. %	≤500 Cases	501 to 1000	1001 to 1500	1501 to 2000	2001 to 3500	Sub-Total
0	2	1				3
0.1–1.0	4	2	4		1	11
1.1–2.0	4	2	1		1	8
2.1–3.0	3	2	1			6
3.1–4.0	4	1				5
4.1–5.0			1			1
5.1–6.0	1	2		1		4
6.1–7.0	2	1				3
7.1–8.0	1					1
8.1–9.0		1				1
9.1–10.0						0
10.1–12.0		1				1
12.1–14.0		2				2
14.1–16.0	1					1
Subtotal	22	15	7	1	2	47

TABLE 6
Number of Laboratories and Frequencies of Pseudomosaicism
(Multiple Cells) Detected

Frequency of Pseudomo. %	≤500 Cases	501 to 1000	1001 to 1500	1501 to 2000	2001 to 2500	2501 to 3000	3001 to 4000	Sub-Total
0	5		1					6
0.1–1.0	9	12	3	1	3	1	3	32
1.1–2.0	5	2						7
2.1–3.0	1	3						4
3.1–4.0								0
4.1–5.0	1		1					2
11.21	1							1
Subtotal	22	17	5	1	3	1	3	52

TABLE 7

Frequency of Trisomic Chromosome in Pseudomosaicism (One or More Cells)

Chromo-some	Cases	%	Chromo-some	Cases	%	Chromo-some	Cases	%
1	9	2.0	9	17	3.8	17	13	2.9
2	101	22.7	10	18	4.1	18	13	2.9
3	12	2.7	11	16	3.6	19	10	2.2
4	8	1.8	12	17	3.8	20	26	5.9
5	21	4.7	13	10	2.2	21	17	3.8
6	12	2.7	14	12	2.7	22	13	2.9
7	24	5.4	15	10	2.2	X	24	5.4
8	17	3.8	16	14	3.2	Y	10	2.2
						Total	444	

TABLE 8

Frequency of Trisomic Chromosome in Pseudomosaicism (Multiple Cells)

Chromo-some	Cases	%	Chromo-some	Cases	%	Chromo-some	Cases	%
1	0		9	5	5.6	17	5	5.6
2	28	31.1	10	4	4.4	18	1	1.1
3	0		11	2	2.2	19	1	1.1
4	1	1.1	12	1	1.1	20	5	5.6
5	3	3.3	13	3	3.3	21	2	2.2
6	1	1.1	14	2	2.2	22	1	1.1
7	10	11.1	15	2	2.2	X	8	8.9
8	3	3.3	16	0		Y	2	2.2
						Total	90	

TABLE 9

Frequency of Monosomic Chromosome in Pseudomosaicism (Multiple Cells)

Chromosome	Cases	%	Chromosome	Cases	%	Chromosome	Cases	%
1	0		9	0		17	6	7.7
2	4	5.1	10	0		18	4	5.1
3	2	2.6	11	1	1.3	19	5	6.4
4	1	1.3	12	0		20	2	2.6
5	0		13	0		21	9	11.5
6	1	1.3	14	1	1.3	22	7	9.0
7	1	1.3	15	1	1.3	X	14	17.9
8	2	2.6	16	3	3.8	Y	14	17.9
						Total	78	

TABLE 10

Cases of Pseudomosaicism Resulting in Live Births with Birth Defects

	Abnormal Karyotype	Outcome
Single Cell Pseudomosaicism	47, XXY	Hydrocephalus
	45, XY, t(9;13)	Multiple congenital anomalies (? Vater's syn.)
	47, XX, +1	Anal atresia
	46, XY, −2, +?, +f	Partial cleft palate
	46, XX, 4q−	Congenital heart disease (deceased)
	47, XX, +2	Aortic stenosis
Multiple Cell Pseudomosaicism	47, XX, +2	Hypospadias
	46, XY, t(1;7)	Club foot
	45, X0 in 46, XY Culture	Microcephaly
	45, X0	Genu recurvatum
	47, XX, +2	Premature with suspected intrauterine growth retardation

TABLE 11

Prenatal Diagnosis of Chromosomal Mosaicism (made from multiple culture vessels)

Karyotype	Phenotype			Abnl.	Confirmation								Confirm.	Source
					Yes			No						
	Nl.	Abnl.	?	Nl.	A.F.	Fet.	L.B.	Nl.	Abnl.	Unsuc.	None	?	Suc. study	
45,X/46,XX n = 21	5F	1 (6F)* (3)	3 (1) 2+	4/11	5	4	4	0	1		4	3	13/14	d,i,m,s, u,v,cc, gg,mm, tt,uu, aaa,3, 11,†
45,X/47,XXX n = 7	2F	(1F)	1 3+	0/3	1	3	1			1	1		5/5	e,m,aaa, bbb,12, 13
46,XX/47,XXX n = 6		(5F)	1+	0/5			5					1	5/5	k,m,mm, qq,zz,14
45,X/46,XrX n = 3	1	1 (1F)	1	1/1	1						1	1	1/1	m,gg,ll
45,X/46,Xi (Xq) n = 2	1F		1	0/1		2							2/2	m,ee
45,X/46,XX/ 46,Xi(Xq) n = 1	1F			0/1			1						1/1	k

96

Karyotype									References
45,X/46,XY n = 17	M9 F1 (M7)	0/17	2	5	1	5	3	8/13	*i,m,q,t,v, y,cc,tt, 11,12, 15,*††
46,XY/ 47,XXY n = 13	M4 (M6)	2 (1)	0/10	4	6	1	2	10/11	*i,l,m,q, aa,gg,ll, bbb,16,* †††
46,XY/ 47,XYY n = 1	1	1/0		1					*oo*
46,XY/ 47,XYY/ 48,XYYY n = 1	1	0/1	1					1/1	*bb*
46,XY/ 46,XYq− n = 1	(1)	0/1	1					1/1	*t*

( )*Liveborn. + Unknown.

†Personal communications: Breg WR, Mahoney J. and Mikkelson M.

††Personal communication: Breg WR, Mahoney J.

†††Personal communications: Doherty R. and Nitowsky H, Schmid R.

TABLE 12
Prenatal Diagnosis of Chromosomal Mosaicism (made from multiple culture vessels)

Karyotype	Phenotype			Abnl. Nl.	Confirmation								Confirm.	Source
					Yes			No						
	Nl.	Abnl.	?	Nl.	A.F.	Fet.	L.B.	Nl.	Abnl.	Unsuc.	None	?	Suc. study	
46/47,+21 n = 20	3 (1)*	6 (2)	2 (1) 5+	8/4	2	10	2	2		1	1	2	14/16	m,q,cc,dd, uu,zz,aaa, bbb,ggg,3, 14,†
46/47,+G n = 1		Down 1		1/0		1							1/1	17
46/47,+20 n = 20	9 (8)	3		3/17		6	0	10		2	1	1	6/16	h,m,q,t, cc,ww, ddd,2,18, 19,20,21, 22,23,††
46/47,+9 n = 6	4	2		2/4	2			3		1			2/5	cc,11,24, †††
46/47,+8 n = 5	3	1	1	1/3		2		2			1		2/4	i,cc,ee,ii, pp
46/47,+18 n = 4	3	1		3/0	1	3							4/4	n,zz, *†
46/47,+13 n = 2	2			2/0		1					1		1/1	m,25

Karyotype				0/1	c
46/47, +7 n = 1	(1)	0/1	1	0/1	
46/45, −9[a] n = 1	(1)	0/1	1	1	*†
46/47, +11 n = 1	1	0/1	1	0/1	tt
46/47, +12 n = 1	1	0/1	1	0/1	*†
46/47, +14 n = 1	(1)	(1)		1	zz
46/47, +15 n = 1	1	0/1	1	1/1	kk
46/47, +16 n = 1	(1)	0/1	1	1	bbb
46/47, +17 n = 1	(1)	0/1	1	0/1	gg
46/45, −17[a] n = 1		1+		1	*†
46/45, −19[a] n = 1	(1)	0/1	1	1	bbb
46/45, −20[a] n = 1		1+		1	bbb

TABLE 12, continued
Prenatal Diagnosis of Chromosomal Mosaicism (made from multiple culture vessels)

Karyotype	Phenotype			Abnl. Nl.	Confirmation								Confirm. Suc. study	Source
	Nl.	Abnl.	?		A.F.	Fet.	L.B.	Nl.	Abnl.	Unsuc.	None	?		
					Yes					No				
46/47, +22 n = 1	1			1/0				1					0/1	zz
46/45, −22ᵃ n = 2		(b)	(1)	1/0			1					1	1/1	*†
46,1q+/ 47, 1q+, +1q n = 1	1			1/0						1				uu
46/46, inv2 n = 1	(1)			0/1				1					0/1	nn
46/46, 4q− n = 1	1			0/1						1				gg
46,r(10)/ 45,−10 n = 1	1			1/0		1							1/1	m
46/46,r(13) n = 1	1			0/1				1					0/1	ff
46/47, +22q− n = 1		(c)		1/0			1						1/1	ff

100

46/45, t(DqDq)/ 45,t(DqGq) n=1	1	0/1	1		0/1 ww
45,t(13q14q)/ 46,t(13q14q), +t(13q14q) n=1	1	0/1	1		0/1 bbb
46/47,+ t(21q21q) n=1	1	1/0	Pl.		0/1 bbb
46/46, t(6p−9p+) n=1	1	0/1	1		0/1 nn
46/45,−20, +t(X;20) n=1	(1)	0/1		1	0 cc
46/46, t(1p−;7q+) n=1	(1)	0/1	Bl.		0/1 t
46/46, t(Xq+; 15q−) n=1	(1)	0/1	Bl.		0/1 t
46/46, t(4q+; 6q−) n=1	(1)	0/1	Bl.		0/1 t

TABLE 12, continued
Prenatal Diagnosis of Chromosomal Mosaicism (made from multiple culture vessels)

Karyotype	Phenotype			Confirmation									Confirm.	
	Nl.	Abnl.	?	Yes				No						
				Abnl./Nl.	A.F.	Fet.	L.B.	Nl.	Abnl.	Unsuc.	None	?	Suc. study	Source
46/46, dup (9)t(9:14) n = 1			1									1		i
46,XX/69,XXX n = 2		1F^d 1^e		2/0		2							2/2	ff,oo

*( ) Liveborn
+Unknown
aMonosomy was detected from two flasks; each flask showing only one abnormal cell.
bNeonatal death
cCat eye syndrome
dSyndactyly
eStillborn
Pl. Placenta
Bl. Blood

†Breg WR, Mahoney J; Rudd N; Solish G; and Styles S. Personal communications.
††Mikhelson M. and Jackson L. Personal communications.
†††Shapiro L, Doherty R. Personal communications.
*† Breg WR, Mahoney J. Personal communication.
*†† Mikkelson M. Personal communication.

TABLE 13
Prenatal Diagnosis of Chromosomal Mosaicism (made from multiple culture vessels)

	Phenotype				Confirmation									
				Abnl.	Yes			No					Confirm.	
Karyotype	Nl.	Abnl.	?	Nl.	A.F.	Fet.	L.B.	Nl.	Abnl.	Unsuc.	None	?	Suc. study	Source
46/47, +mar. n = 21*	6	2 (6)** (1)	3 (3)	3/12	1	9	3	3		5			13/16	c,g,l,m, q,t,ff,ss, aaa,bbb, ddd,eee, 24,†
47, +mar./ 48, +mar., +mar. n = 1*		(1)		0/1							1			i
47, +mar./ 48, +mar., +mar./49, +mar.,+mar., +mar. n = 1	1			0/1						1				uu

*In three cases, one parent is mosaic for an identical marker chromosome. ** ( ) Liveborn.
† Breg WR, Mahoney J. Personal communication.

TABLE 14

Summary of Prenatal Diagnosis of True Chromosome Mosaicism*

Type of Mosaicism	Total Number	No. abnl. pheno./ Total exam.	(%)	No. cytogen. confirmed (abort. & liveb.)/ Successful studies	(%)
Sex chromo.	73	6/57	(10.5)	46/53	(86.8)
Autosomes	89	28/74	(37.8)	34/65	(52.3)
Mar. chromo.	23	3/17	(17.6)	13/16	(81.3)
Total	185	37/148	(25.0)	93/134	(69.4)

*All diagnoses were based upon multiple culture vessels.

REFERENCES

1. Hamerton JL, Boue A, Cohen MM, et al. Prenatal diagnosis: Past, present, and future. Section 2: Chromosome disease. Prenatal Diagnosis (Special Issue) Dec. 1980; 11-21.
2. Boue J, Nicolas H, Barichard F, Boue A. Le clonage des cellules du liquide amniotique, aide dans l'interpretation des mosaiques chromosomiques en diagnostic prenatal. Ann Genet 1979; 22: 3-9.
3. Peakman DC, Moreton MF, Corn BJ, Robinson A. Chromosomal mosaicism in amniotic fluid cell cultures. Am J Hum Genet 1979; 31: 149-55.
4. Friedrich U, Nielsen J. Chromosome studies in 5,049 consecutive newborn children. Clin Genet 1973; 4: 333-43.
5. Jacobs PA, Melville M, Ratcliffe S. A cytogenetic survey of 11,680 newborn infants. Ann Hum Genet 1974; 37: 359-76.
6. Hamerton JL, Canning N, Ray M, Smith S. A cytogenetic survey of 14,069 newborn infants. Clin Genet 1975; 8: 233-43.
7. Higurashi M, Iijima K, Ishikawa N, Hoshina H, Watanabe N. Incidence of major chromosome aberrations in 12,319 newborn infants in Tokyo. Human Genet 1979; 46: 163-72.
8. Hsu LYF. Prenatal cytogenetic diagnosis: a mini-review. In: Young, B. ed. Perinatal medicine today. New York: Alan R. Liss, 1980: 3-25.
9. Summit RL, Breg WR, Hook EB, et al. Spurious cells in amniotic cell cultures from multiple laboratories. Am J Hum Genet 1980; 32: 90A.

10. Cantu JM, Del Castillo V, Kopelivich-Chametz M, *et al.* The pregnancy in women over 40 years of age. A gyneco-obstetric, pediatric, cytogenetic and socio-economic evaluation. Birth Defects Conference. Kansas City: March of Dimes Foundation. 1975: 219.

11. Hsu LYF, Kaffe S, Yahr F, *et al.* Prenatal cytogenetic diagnosis: First 1,000 successful cases. *Am J Med Genet* 1978; 2: 365-83.

12. Kardon N, Krauss M, Davis J, Jenkins E.   Chromosome mosaicism in amniotic fluid cell cultures. *J Ped* 1977; 90(3): 501-2.

13. Kohn G, Cohen MM, Beyth Y, Ornoy A. Prenatal diagnosis and gonadal findings in X/XXX mosaicism. *J Med Genet* 1977; 14: 120-3.

14. King CR, Magenis RE, Prescott G, Bennett S, Olsen C. Mosaicism and non-modal cells in amniotic fluid cultures. *Am J Hum Genet* 1978; 30: 85A.

15. Sutherland GR, Bowser-Riley SM, Bain AD. Chromosomal mosaicism in amniotic fluid cell cultures. *Clin Genet* 1975; 7: 400-4.

16. Milunsky A. Prenatal diagnosis of chromosomal mosaicism. *J Ped* 1976; 88: 365-66.

17. Simola K, Aula P, Ryynanon M, von Koskull H. Incomplete prenatal diagnosis of G-trisomy mosaicism. *Clin Genet* 1978; 13: 500-3.

18. Kardon N, Lieber E, Davis J, Hsu LYF. Prenatal diagnosis of trisomy 20 mosaicism. *Clin Genet* 1979; 15:267-72.

19. Rudd NL, Gardner HA, Worton RG. Mosaicism in amniotic fluid cell cultures. *Birth Defects* 1976; 13,3D: 249-88.

20. Rodriguez ML, Luthy D, Hall JG, Norwood TH, Hoehn H. Amniotic fluid cell mosaicism for presumptive trisomy 20. *Clin Genet* 1978; 13: 164-68.

21. Nevin NC, Nevin J, Thompson W. Trisomy 20 mosaicism in amniotic fluid cell culture. *Clin Genet* 1979; 14: 440-43.

22. McDermott A, Gardner A, Ray B. Trisomy 20 mosaicism in a fetus. *Pre natal Diagnosis* 1981; 1: 147-51.

23. Nisani R, Rappaport S, Felsenburg T, Chemke J. Trisomy 20 mosaicism in amniotic fluid cell cultures. Poster Session 19–16, Abstracts of the 6th International Congress of Human Genetics. Jerusalem, Israel, Sept. 13-18, 1981: 301.

24. Polani PE, Alberman E, Alexander BJ, *et al.* Sixteen years experience of counseling, diagnosis and prenatal detection in one genetic center: progress, results, and problems. *J Med Genet* 1979; 16: 166-75.

25. Bloom AD, Schmickel R, Barr M, Burdi AR. Prenatal detection of autosomal mosaicism. *J Ped* 1974; 84: 732-34.

26. Hook EB. Exclusion of chromosome mosaicism: Tables of 90%, 95% and 99% confidence limits and comments on use. *Am J Hum Genet* 1977; 29: 94-7.

WHO GETS AMNIOCENTESIS?

Barbara A. Bernhardt

R. M. Bannerman

INTRODUCTION

This slightly provocative title is used to introduce some simple inquiries into the question, "Who gets amniocentesis?" We will not be discussing the indications for amniocentesis, which are well recognized. Instead, we will discuss some factors which may influence amniocentesis utilization or availability. The question deserves critical consideration because the answers may influence planning and utilization for cytogenetic laboratories and genetic units, and have important effects on the incidence of chromosome disorders.

Overall, medical geneticists, most physicians, and much of the public approve and support the use of genetic amniocentesis in high risk situations.[1,2] Although clearly increasing from year to year, recorded utilization of amniocentesis is generally lower than might be expected.[3,4,5]

UTILIZATION

Amniocentesis utilization rates can only be computed for amniocentesis based on the indication of advanced maternal age, because the denominator for women obtaining it for other indications (*e.g.* previous child with trisomy, previous child with a neural tube defect) is unknown. Thus, rates are expressed as the percentage of women who receive amniocentesis among those giving birth at or above a certain age, or "age-eligible" women. Estimates from various places are available and show marked variations. In the West Midlands of England in 1977, 16% of mothers over 40 received amniocentesis, but there was marked local variation ranging from over 50% in the city of Birmingham to apparently zero in Burton.[6] In a study in the USA in 1978 there was a fourfold range of variation in utilization in different parts of the country with 6.7% and 28.9% of pregnant women 35 years of age and older in Nebraska and Manhattan, repectively, obtaining amniocentesis.[5]

CLINICAL GENETICS: PROBLEMS IN
DIAGNOSIS AND COUNSELING

In the state of New York, data are available from the New York State Chromosome Registry at Albany. It is likely that nearly all amniocenteses in the state were reported to the registry or otherwise ascertained; results from 1979 are available by county.[7] Not surprisingly, the highest rates are in metropolitan areas, in and around New York City, Rochester and a few other places. The micro-geographic effects deserve examination. The present study concentrates on the Western New York area, Health Systems Agency Region 1, comprising the eight counties surrounding Buffalo (Figure 1) with a population of nearly two million.

FACTORS DETERMINING UTILIZATION

Four steps leading to amniocentesis may be considered.

1. The woman is made aware that the technique exists and may be relevant to her situation.
2. The opportunity to utilize the technique is made available.
3. The woman chooses to accept it.
4. Amniocentesis is successfully performed.

Many factors influence the outcome at each step and determine whether or not a woman eventually gets amniocentesis. Factors can be divided into those relating mostly to the mother, and those relating to the obstetrician.

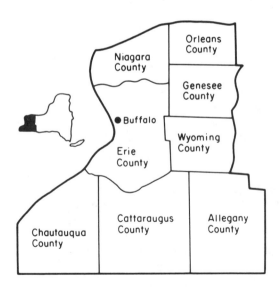

Fig. 1. The eight western counties comprising the New York Health Systems Agency Region 1.

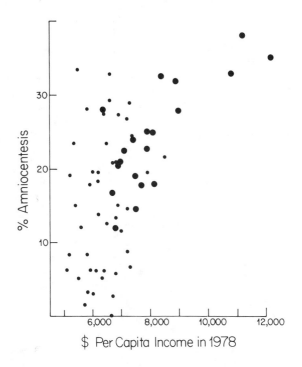

$ Per Capita Income in 1978

Fig. 2. Scatter diagram showing percent utilization of amniocentesis by county of New York State[7] compared with per capita income by county in 1978.[8]

The larger symbols represent counties where there were more than 100 age-eligible women delivering in that county during 1979; the smaller symbols represent counties where there were less than 100 such women. The correlation coefficient, r= 0.736 (P<0.01), is calculated for the counties with larger numbers of women only.

MATERNAL FACTORS

Cost and Income

In some countries the total cost of amniocentesis is borne directly by a nationalized health service and, therefore, only indirectly by individual patients. In the United States, there is great variability in the extent to which the procedure and the associated counseling, ultrasound and laboratory work are covered by third party payers or other agencies, through subsidy by various federal and other grant sources, and by the patient.

TABLE 1
Education of Erie County Women Obtaining
Amniocentesis
(Median years; age matched, 35 and older)

Years	Total county	Amnio group
1974-76	12.1	13.9
1977-78	12.3	13.4
Total	12.2	13.6

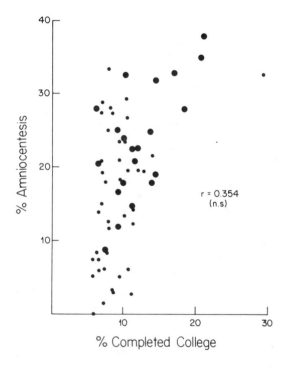

% Completed College

Fig. 3. Scatter diagram showing percent utilization of amniocentesis by county of New York State[7] compared with percent of persons over 25 having completed college, as an index of education.[10]

The larger symbols represent counties where there were more than 100 age-eligible women delivering in that county during 1979; the smaller symbols represent counties where there were less than 100 such women. The correlation coefficient, r=0.354, is not significant but the trend is evident. The point to the extreme right is that for Tompkins County, see text.

The New York State data show an obvious trend in the direction of greater amniocentesis utilization with greater income. In Figure 2, the utilization percentage for each county of the state is plotted against per capita income as taken from the Department of Commerce data.[8] There is a marked and significant correlation between utilization and income. The highest utilization rate reported in the state is for Westchester county (37.9%), one of the most affluent counties. Exceptions are mostly in the direction of high utilization rates in lower income areas, such as in Tompkins County where low per capita income in a thinly populated area is possibly outweighed by having a relatively large group of educated and informed people associated with Cornell University in Ithaca.

Education

In earlier studies, we noted that women having amniocentesis tended to be a highly educated group. We thought this was partly because they were better informed, and a higher proportion of these women were pursuing professional careers and, therefore, tended to start their families at older ages.[9]

These data have recently been updated for Erie County by comparing median years of education of women obtaining amniocentesis because of age, with that of all similarly-aged women in the county delivering babies from 1974-78 (Table 1). In the total, there is more than one year difference in years of education for amniocentesis mothers. When the period of observation is broken up, there appears to be a trend towards reducing the difference, *i.e.* a suggestion that as utilization increases, more women in the group with less education are receiving amniocentesis.

To examine the role of education on a statewide basis, we have used the percentage of persons over 25 years having a college degree, obtained from census data[10] as an index. Figure 3 compares amniocentesis utilization rates by county with the percent of persons who have completed college. The correlation coefficient is low and not statistically significant, but a trend can be discerned toward greater amniocentesis utilization in counties where more people have completed college. In comparison, Tomkins County appears as the "most educated", 28.4% of persons over 25 having completed college.

Women's Religion

Data on the religion of 308 Western New York women receiving amniocentesis from 1974-78 are presented in Table 2. In the group studied, the proportion of women of various religions receiving amniocentesis is approximately the same as that of the community in general. Thus, 42-45% of women receiving amniocentesis were Catholic, while 46% of the general population of Western New York is reported to be Catholic.

TABLE 2
Religions of Women Having Amniocentesis

	1974-76 (%)	1977-78 (%)
Protestant	51	49
Roman Catholic	42	45
Jewish	5	1
Other/none	2	5

TABLE 3
The Pro-Life Bonus
(Western New York, 297 women, 1974-78)
In the absence of amniocentesis, women with usual
indications for the procedure, would have:

	n	%
Carried on	243	82.0
Avoided pregnancy	25	8.4
Terminated pregnancy	12	4.0
Artificial insemination by donor	1	0.3
Uncertain	16	5.3

TABLE 4
Initial Source of Information (1974-78)

Source	Education	
	High School or less	College or more
Physician	81%	49.4%
Media	16%	30.2%
Friend/relation	3%	5.2%
At work	0	11.6%
School	0	3.5%

Available Alternatives

Given that amniocentesis is generally an available option in high risk pregnancies, have women modified their decision to undertake or continue a pregnancy because of its availability? As a part of our follow-up we retrospectively asked women what they might have done if amniocentesis had not been available to them. While this is clearly a speculative inquiry, the results (Table 3), which we previously reported,[11] are of importance to the theme. While it appears that 82% of women would have carried on without amniocentesis, a total of 12.4% would either have not undertaken the pregnancy or would have had an abortion. Against this must be set the approximately 3% of monitored pregnancies where the diagnosis of a fetal abnormality led to therapeutic abortion. We are left with the possibility of 9% of pregnancies saved, which is a pro-life bonus of amniocentesis.

Sources of Information

In the same survey of women obtaining amniocentesis, we inquired about the initial source of information about the procedure. The responses were examined by comparing the source of information with the completed years of education (Table 4). The majority of women with a high school or less education learned of the procedure from a physician, usually their obstetrician. Half of the women with more years of education learned about amniocentesis from other sources, especially the media and at work. With obstetricians as the major source of information, we turned to examining characteristics of obstetricians in relation to amniocentesis utilization.

OBSTETRICIAN FACTORS

In our area, we compared certain obstetrician characteristics with the extent to which individual obstetricians referred patients for amniocentesis.[12] The area lends itself to study because all amniotic fluid analyses for the eight Western New York counties are carried out in the Cytogenetics Laboratory of Buffalo Children's Hospital. This area had an overall utilization rate of amniocentesis by pregnant women 35 years and older of 14.2% in 1979,[7] and a lower one in the five years from 1974-78.

Similar observations were made by questionaire and telephone inquiries sent to obstetricians in Montreal in 1977-78 in the study by Lippman-Hand and Cohen.[13] Our study differs because we have recorded observed referrals or non-referrals and did not directly survey individual obstetricians.

Overall Referral Rates

The 173 obstetricians in our area were categorized as to whether they had carried out mid-trimester amniocentesis or referred patients for it: none, one or two, or three or more times in the five year study period (Table 5). It emerged that 47% of area obstetricians had *not* referred any patients during the entire period, and only 24% referred three or more women for amniocentesis.

The three groups of obstetricians were examined according to five independent variables: board certification, hospital affiliation, practice location, age, and religion. Most of the information about these characteristics came from the Health Systems Agency.

Board Status and Hospital Affiliation

Board certification status was significantly related to referral (Table 6). Sixty-three percent of the non-board certified obstetricians had not made amniocentesis referrals versus 39% of those with certification (X^2_{1df} = 9.22, P<.005). Board certified obstetricians were more likely to refer three or more women (28%) than those without certification (18%).

Hospital affiliation was also significantly related to referral: 65% of the obstetricians affiliated with a teaching hospital made one or more referrals versus 46% who were affiliated with non-teaching hospitals (X^2_{1df} = 5.52, P<.025). The proportion of obstetricians referring three or more women for amniocentesis was higher for those with teaching hospital affiliations (36%) than for those without (18%).

Location of Practice

Obstetricians were classified as urban if they practiced in the metropolitan counties of Erie and Niagara; rural, if they practiced in the other six counties. During the study period, 53% of rural obstetricians made no referrals as apposed to 46% of urban obstetricians (X^2_{1df} = .53, P<.750); the difference is not statistically significant.

Age

The difference in age between referring and non-referring obstetricians is not statistically significant (t = 1.51, .10<P<.20), but there is an evident trend. (Table 7) Furthermore, the mean age of obstetricians referring three or more patients is significantly lower than the mean age of those referring fewer than three patients (t = 3.45, P<.001).

Religion

During the period, no amniocentesis referrals were made by 34 out of 70 Catholic obstetricians (49%), 12 out of 34 Protestant obstetricians (35%), five out of 13 Jewish obstetricians (39%) and six out of 22 other obstetricians (27%) (Table 8). The proportion of obstetricians who are Catholic is approximately the same as the proportion of total population that is Catholic, just under 50%, but they make a smaller contribution to referrals for amniocentesis. If the proportion of referring obstetricians is compared between Catholic and non-Catholic obstetricians, the religion of referring obstetricians is not statistically significant (X^2_{1df} = 3.34, .05<P<.10). However, the pro-

TABLE 5
Obstetricians: Amniocentesis Referrals
(Western New York; 1974-78)

No referrals	82	(47.4%)
1–2 referrals	49	(28.3%)
3 or more	42	(24.3%)
	173	

TABLE 6
Obstetricians' Board Status and Hospital Affiliation

	Certified (%)	Not certified (%)	Teaching hospital (%)	Non-teaching hospital (%)
No referrals	39	63	35	54
1–2 referrals	33	19	29	28
3 or more	28	18	36	18

TABLE 7
Obstetricians' Ages
(Western New York; 1974-78)

	Mean Age
No referrals	52.4
1–2 referrals	52.3
3 or more	47.5

TABLE 8
Obstetricians' Religion (Western New York; 1974-78)

	Catholic (%)	Protestant (%)	Jewish (%)	Other (%)
No referrals	49	35	27	41
1–2 referrals	31	35	32	31
3 or more	20	30	41	28

portion of Catholic obstetricians referring three or more patients is smaller than that of other obstetricians.

DISCUSSION AND CONCLUSION

The question "Who gets amniocentesis?" is important. We have tried to go a little way towards answering it. Among the factors examined in New York State, there is significant correlation of amniocentesis utilization with per capita income by county, and a less marked correlation with education. In the western area of the state we confirmed that the number of years of education is positively related to amniocentesis utilization, but this is becoming less marked with time as utilization increases. Religion of the women did not appear to be an important factor.

What emerges strikingly is the apparent role of obstetrician factors. It is from their obstetricians that most women, particularly those with less years of education, first learn of the procedure, yet in the area studied nearly half of the obstetricians had not referred a patient for amniocentesis at all during the five year study period. Other factors associated with less or no referrals include the absence of Board Certification, Catholic religion, and older age. We do not know at what level these factors operate, whether at the level of informing the patient versus not informing her, or acceptance by the patient, or both. If greater use of amniocentesis is desired, and many feel that it is, there is clearly a need for continuing professional education for obstetricians and others.[14] There is also a need to continue to analyze periodically the factors which influence utilization in different areas.

ACKNOWLEDGMENTS

Studies were supported in part by the U. S. Department of Health and Human Services Bureau of Maternal and Child Health, Human Genetic Program Project 417, a contract from the New York State Birth Defects Institute, and a grant from the March of Dimes Birth Defects Foundation.

REFERENCES

1. Budney S. Attitudes of 40-year-old college graduates towards amniocentesis. *Brit Med J* 1975; 2: 1475-77.
2. Sell RR, Roghmann KJ, Dougherty RA. Attitudes toward abortion and prenatal diagnosis of fetal abnormalities: Implications for educational programs. *Soc Biol* 1978; 25: 288-301.
3. Goldstein AL, Dumars KW. Minimizing the risk of amniocentesis for prenatal diagnosis. *JAMA* 1977; 238: 1336-38.
4. Ferguson-Smith MA. Human cytogenetics. *In*: Birth Defects: Proceedings of the Fifth International Conference. Amsterdam: Excerpta Medica, 1978.

5. Adama MM, Finley S, Hansen H, *et al.* Utilization of prenatal diagnosis in women 35 years of age and older in the United States, 1977-1978. *Amer J Obstet Gyneco* 1981; 139: 673-77.

6. Edwards JH, Well T. Amniocentesis rates in older women. *Lancet* 1979: 2: 1301.

7. Hook EB, Schreinemachers D, Cross PK. Prenatal cytogenetic diagnosis in New York State in 1979: A report from the New York State Chromosome Registry. Appendix. Vol. 25Y. New York State Chromosome Registry, 1981.

8. Grainger WT, Sheftic RJ. Personal income in areas and counties of New York State. 1978. State of New York Department of Commerce. Bureau of Business Research Bulletin no. 48, August 1980.

9. Bannerman RM, Gillick D, Van Coevering R, *et al.* Amniocentesis and educational attainment. *New Eng J Med* 1977; 297: 449.

10. U.S. Bureau of the Census. County and City Data Book, 1972. Washington: U.S. Government Printing Office, 1973.

11. Bernhardt BA, Bannerman RM. The pro-life bonus of amniocentesis. *New Eng J Med* 1980; 302: 925.

12. Bernhardt BA, Bannerman RM. The influence of obstetricians on the utilization of amniocentesis. *Prenatal Diagnosis.* In press.

13. Lippman-Hand A, Cohen DI. Influence of obstetricians' attitudes on their use of prenatal diagnosis for the detection of Down's syndrome. *Can Med Assoc J* 1980; 122: 1381-86.

14. Simpson JL, Elias S, Gatlin M, *et al.* Genetic counseling and genetic services in obstetrics and gynecology: Implications for educational goals and clinical practice. *Amer J Obstet Gynecol* 1981; 140: 70-80.

INTERPRETATION OF RECENT DATA PERTINENT TO GENETIC COUNSELING FOR DOWN SYNDROME: MATERNAL-AGE-SPECIFIC-RATES, TEMPORAL TRENDS, ADJUSTMENTS FOR PATERNAL AGE, RECURRENCE RISKS, RISKS AFTER OTHER CYTOGENETIC ABNORMALITIES, RECURRENCE RISK AFTER REMARRIAGE

Ernest B. Hook
Philip K. Cross

INTRODUCTION

Over the past few years our group in the Birth Defects Institute has undertaken, with collaborators, extensive regression analyses of rates of Down syndrome by one year maternal age intervals.[1-6] Rates from these studies appear in Tables 1 and 2. One of our initial hopes was that this would enable more precise genetic counseling than that based on previous analyses by five year maternal age intervals. The derived rates in our studies fit well the observed data but, because of observations of a possible temporal change in live birth rates[7-9] and concern about cases of paternal origin,[10] questions have arisen as to how appropriate the rates are for counseling and what adjustments should be undertaken for factors such as paternal age. In this paper we address such adjustments in detail, and also consider other factors pertinent to the occurrence of a Down syndrome live birth that may alter risk figures derived from large population studies.

RATES ARE NOT NECESSARILY RISKS

First, it is important to distinguish between a "rate" and a "risk". As the term is usually used in human genetics, a rate is the proportion of events (*e.g.* chromosome abnormalities) among some outcomes (*e.g.* live births) derived from a study or studies of a specific population during some time period. A risk is an individual's probability of such an event in the future. Age-specific-rates may be thought of as "past tense"; risks as "future tense". A rate schedule from one study may or may not provide an individual's precise risk at some time in the future.*

*This distinction is occasionally not appreciated, particularly by some epidemiologists who, in an unqualified way, regard rates as average risks. See Ref. 11.

CLINICAL GENETICS: PROBLEMS IN
DIAGNOSIS AND COUNSELING

119

TABLE 1
Observed (and Projected) Maternal Age Specific Rates of Down Syndrome per 1,000 Live Births

Maternal age	Upstate NY 1964-74*[1]	Mass. 1958-65[2]	Sweden 1968-70[3]	Estimated range** L	U	Amniocentesis rate+	80% of amniocentesis rate+
21	0.6	0.6	0.7	0.6	0.7	—	—
23	0.7	0.7	0.8	0.7	0.8	—	—
25	0.8	0.7	0.8	0.7	0.9	—	—
27	1.0	0.8	0.9	0.8	1.0	—	—
29	1.1	0.9	1.0	0.8	1.2	—	—
31	1.2	0.9	1.1	0.9	1.3	—	—
33	1.7	1.6	1.5	1.5	1.9	—	—
35	2.7	2.5	2.5	2.5	3.9	3.7	2.9
36	3.5	3.2	3.3	3.2	5.0	4.7	3.8
37	4.4	4.1	4.3	4.1	6.4	6.1	4.9
38	5.7	5.2	5.6	5.2	8.1	7.9	6.3
39	7.2	6.7	7.3	6.7	10.5	10.1	8.1
40	9.2	8.5	9.6	8.5	13.7	13.1	10.5
41	11.7	10.9	12.5	10.9	17.9	16.9	13.5
42	14.9	13.8	16.4	13.8	23.4	21.8	17.4
43	19.0	17.7	21.4	17.7	30.6	28.1	22.5
44	24.2	22.5	28.0	22.5	40.0	36.2	29.0
45	30.8	28.7	36.6	28.7	52.3	46.8	37.4
46	39.3	36.6	47.8	36.6	68.3	60.3	48.3
47	50.0	46.7	62.5	46.7	89.3	77.8	62.3
48	63.8	59.5	81.7	59.5	116.8	100.4	80.3
49	81.2	75.9	106.8	75.9	152.7	129.6	103.7

*Original published estimates. For revised estimates using this source see next table.   +See addendum A
**Adjusted for temporal increase.
†Rates at amniocentesis for maternal ages 35 and up were calculated from data on 29,229 prenatal diagnoses obtained by pooling the results from six series. The analysis was restricted to women studied for elevated parental age or a reason other than a known cytogenetic risk. See *Amer J Hum Genet* 1980; 33: 120A. The rates have been adjusted to maternal age at expec-

TABLE 2
Rates of Down Syndrome per 1,000 Live Births Using Birth Certificates [++]

Maternal Age	Upstate NY 1963-74*	Upstate NY (Revised) 1963-74**	Upstate NY 1963-69†	Upstate NY 1970-74††	Ohio 1970-74+
21	0.6	0.7	0.7	0.8	0.7
23	0.7	0.8	0.7	0.8	0.8
25	0.8	0.8	0.8	0.9	0.8
27	1.0	1.0	0.9	1.0	0.9
29	1.1	1.2	1.0	1.3	1.1
31	1.2	1.5	1.3	1.7	1.4
33	1.7	2.1	1.8	2.4	2.0
35	2.7	3.1	2.6	3.7	3.1
36	3.5	3.8	3.3	4.6	4.0
37	4.4	4.8	4.1	5.8	5.1
38	5.7	6.1	5.2	7.4	6.7
39	7.2	7.8	6.7	9.5	8.8
40	9.2	10.0	8.6	12.3	11.6
41	11.7	12.9	11.1	16.0	15.4
42	14.9	16.8	14.4	20.7	20.6
43	19.0	21.8	18.8	27.0	27.7
44	24.2	28.5	24.6	35.3	37.2
45	30.8	37.2	32.3	46.2	50.1
46	39.3	48.6	42.3	60.4	67.5
47	50.0	63.6	55.6	79.2	91.0
48	63.8	83.3	73.2	103.8	122.9
49	81.2	109.2	96.3	136.1	166.0

*Original estimates[1]. The equations used were: for maternal ages 20 to 30: $(1000)y = .02258x - .25498$, and for maternal 33 to 49: $(1000)y = (\exp [.24211x - 8.44711])$ where y = rate and x = maternal age. The correction factor for underreporting was 0.375.

**Based on a more attractive regression model ($y = .0023 + \exp [-16.67338 + .27270x]$) and more appropriate correction factor for underreporting, 0.336.

†Regression equation: $y = .00025 + \exp (-16.9038 + .2769x)$; correction factor for underreporting = 0.374.

††Regression equation: $y = .00021 + \exp (-16.53561 + .27283x)$; correction factor for underreporting = 0.311.

+Regression equation (including three cases detected by prenatal diagnosis) $y = .00024 + \exp (-17.66192 + 0.30217x)$; correction factor for underreporting = 0.348.

(All studies are for white population only.)
++See addendum B

Often we have no reason *not* to extrapolate rates to individuals in other populations and future time periods, and thus use them as risk schedules. Geneticists do this routinely. For example, for many multifactorial defects American geneticists use figures on recurrence rates derived from studies in the United Kingdom to counsel individuals, often of different ethnic backgrounds and living 3,000–6,000 miles away, on the risk of events occurring at some indefinite time in the future. This is done because it is assumed, *in the absence of evidence to the contrary*, that the rates of conditions involved are stable. But all genetic counselors, it is hoped, also recognize that if the live birth prevalence rate of a trait varies significantly temporally, geographically, or ethnically, as for example neural tube defects do, then one must be circumspect about using data from one population to counsel individuals from another.

With regard to Down syndrome, despite a great number of claims of seasonality and other variability (see discussion in reference 12) genetic counselors, as far as we are aware, assume that (maternal age specific) live birth prevalence rates are relatively stable over time and among populations. Indeed, aside from maternal age and some rare factors (previous trisomic offspring, parental mosaicism, parental 21-translocation carriers, presence of affected relatives) no other factors have been shown to be *unequivocally* correlated with live birth prevalence rates, despite many suggestions to the contrary.[12] Nevertheless, one cannot exclude the possibility of temporal and spatial variation in maternal age specific rates not attributable to selective changes in abortion. Indeed, two recent reports have noted evidence for a temporal increase in live birth prevalence rates. (There is also a very suggestive report of higher maternal age specific rates in Asiatic and North African Jews than in European or North American groups such as those on which data are presented in Table 1. This difference, however, must still be confirmed by further study. With this possible exception, we are aware of no good evidence for ethnic variation in rates. See further discussion in references 12, 13 and 14.)

POSSIBLE TEMPORAL CHANGES IN AGE ADJUSTED LIVE BIRTH PREVALENCE RATES FOR DOWN SYNDROME

Observations of Evans *et al*,[7] analyzed in detail elsewhere,[12] indicate an increase of about 60–70% in rates of those with Down syndrome born to women 35 and over occurred in 1970-74, compared to 1965-69 in Manitoba. We had seen little change in crude or age-adjusted rates for Upstate New York, but prompted by their report, we analyzed specifically the data for those 35 and over and found an increase in the rate in 1970-74 compared to 1965-69, albeit of lesser magnitude than the increase in Manitoba.[8,9] Regression derived rates are higher in the latter interval at all ages because of this (Table 2). We cannot find any statistical artifact that can explain the trend. It should be noted, however, that while there are suggestive trends in Montreal,[15] one county in U.K.,[16] and Denmark,[17] consistent with this change, data from British Columbia,*[17] Israel, and Australia[19] suggest no such change, and the trend may result from statistical fluctuation.

What are the implications of this observation for genetic counseling for the risk of Down syndrome? The published live birth rates are for the most part based on data obtained before this apparent change. One course is to cite a range in rates from the lowest in any of the earlier studies to the highest at any age, with additional adjustment in those 35 and over for the apparent temporal change of about 30% noted in the New York State data. (See column 5, Table 1.) We are confident that best risks for individuals today without *other* known risk factors lie in the ranges given at each age, but of course have no certainty of where within the ranges they lie. Nevertheless, data from amniocentesis provide some reassurance that the data of the earlier live birth studies may still be appropriate, and that the upper limit of the range cited in column 5, Table 1 is unnecessarily high.

If we pool all of the amniocentesis data (excluding those with risk factors aside from maternal age) and do a regression analysis upon these data,** the derived rates, adjusted for maternal age at time of expected live birth, are as given in column 6, Table 1. If one assumes that (a) 20% of the Down syndrome cases diagnosed prenatally would not survive in live births[20] and (b) such loss is independent of maternal age, then the regression derived maternal age specific rates at time of expected live birth (column 7, Table 1) are almost exactly equal to those given in some earlier publication, *e.g.* in the Swedish study on live births in years 1968-70 (column 4, Table 1). This provides some reassurance that the use of such live birth data will not result in gross underestimates of live birth rates. (If Polani *et al* are correct, data from amniocentesis on those studied ostensibly only because of advanced maternal age are biased upwards because of inclusion of those with putative cryptic factors that elevate the true risks over those associated with age alone.[21] If this is the case then live birth rates derived from such studies at amniocentesis, diminished by 20% for spontaneous fetal loss after amniocentesis as in column 7, Table 1, would provide only an *upper boundary* of expected live birth risk, providing even greater reassurance that the Swedish live birth rates, which are about equal to these putative upper limits, are not *underestimates.*[+]

For this reason, at the present time we believe that the most appropriate rates for counseling are those derived from the Swedish study, or from amniocenteses studies decremented by about 20–30%. Sufficient data should be available in the near future to investigate spatial and temporal variation in

---

*Ward R. Personal communication.   +See also addendum A
**Schreinemachers DM, Cross PK, Hook EB. Rates of trisomy 21, 18, 13 and other chromosome abnormalities in 29,229 amniocenteses compared to estimated rates in live births. (See also *Amer J Hum Genet* 1980; 33: 120A.) Submitted for publication.

amniocentesis rates. However, we still do not have an adequate explanation for the apparent temporal changes in live birth rates in 1970-74.*

TRENDS IN INTERCHANGE TRISOMY DOWN SYNDROME

Most of the Down syndrome phenotype, 95–97%, is attributable to 47,+21 so any major fluctuations in the phenotypes are almost certainly due to changes in the 47,+21 genotype.[12] Some fluctuation has occurred, however, in the proportion of cases attributable to mutant interchange trisomic cases, *i.e.* those associated with 46 chromosomes and unbalanced G/21 or D/21 Robertsonian translocation.[23-25] While these changes appear to indicate an increase in mutation rates for those with interchange trisomic Down syndrome in 1973-75 (at least as measured in live births) they do not appear to have major implications for genetic counseling. A summary of changes in the proportions appears in Table 3, and of live birth prevalence rates in Table 4. (Extensive data on the proportions themselves for Down and Patau syndrome appear in tables in the appendix.)

---

*There is some variability within the live birth studies presented in Tables 1 and 2 that warrants further comment. Assuming temporal and geographical differences are not significant, we believe it is *most likely* that the Swedish data are representative of "true" live birth rates, and that the Massachusetts data involve about 10% underascertainment of cases. Similarly we suspect a similar or greater magnitude of underascertainment has also occurred in the British Columbia[22] and Australia[19] studies. With regard to birth certificate studies presented in Table 2, the issue is somewhat more complicated, because these estimates are all derived using a correction for underascertainment that has a sampling fluctuation of its own. The original published rates[1] used a crude regression model, and were based on a completeness of 37.5%. The collection of additional data, and further refinement of estimate of completeness have revised this figure downwards to 33.6%. Estimates using this figure and a more refined regression model as well appear in column 3, Table 2, and are somewhat higher than those previously published. In addition, estimates for the intervals 1963-69 and 1970-74 derived using separate completeness estimates for each interval also appear in Table 2, as do data from a 1970-74 birth certificate study in Ohio. It is noteworthy both estimates from the birth certificate studies for 1970-74 are notably higher than the Swedish rates and, at upper extremes of maternal age, are similar to the rates at amniocentesis which were derived from studies in 1975-81. In any event, the estimates from amniocentesis, decremented appropriately, suggest that whatever the final explanation of the apparent high rates in 1970-74 interval noted in Table 2 may be, the Swedish live birth rates appear appropriate for counseling now.

See also addendum B

TABLE 3
Proportion of Interchange Down Syndrome Known to be Mutant
Among All With Down Syndrome Phenotype by Year of Birth*

1968	1969	1970	1971	1972	1973	1974
4.2%	3.0%	2.0%	2.6%	2.4%	4.9%	5.4%

1975	1976	1977	1978	1979	1980
4.1%	2.3%	4.8%	2.4%	1.1%	5.6%

*See appendix for more extensive data by maternal age, and cytogenetic sub-
type. Data are analyzed on reports, received by April 15, 1981.

TABLE 4
Trends in Estimated Prevalence Rates of Interchange Trisomic Down
Syndrome per 100,000 Live Births Attributable to *De Novo* Mutation
in New York State, 1968-1977*

1968	1969	1970	1971	1972	1973	1974
4.0	4.0	3.4	3.2	3.4	7.0	7.0

		1975	1976	1977		
		6.2	4.4	6.0		

*Derived from most likely estimate of mutation rates based on data reported
to the New York State Chromosome Registry by June 15, 1978.[25]

PATERNAL AGE

In 1933 Penrose[26] and Jenkins[27] showed that the predominant explanation for the parental age effect in Down syndrome was maternal age, but
these workers could not exclude a slight effect of paternal age as well.[28] In
the absence of further evidence, many subsequently concluded that in fact
maternal age *only* was responsible for the parental age association. However,
the recent discovery that 20–25% of cases are of paternal origin[10] has led
to the expectation by many that there must be a significant positive paternal
age effect as well as maternal age effect. This may be an erroneous expectation.
For example, for the XYY genotype, in which the extra chromosome is known
to be of paternal origin, a slightly *negative* association with paternal age has
been reported.[29] We believe there can be no *a priori* expectation of a paternal

age effect for Down syndrome despite known paternal origin of the extra 21 chromosome and effects must be demonstrated by actual evidence, not by extrapolation.

There are several recent studies of a possible association of paternal age and Down syndrome. Some of these were positive and claimed "statistical significance,"*[30-34] some showed suggestive albeit non-significant trends,[35] and some were negative.[33,35-38] (Some reports included both positive and negative studies.[33,35]) The most dramatic effect has been claimed by Stene et al:[30] a two fold excess for men 55 and over after purported control on maternal age. This has been criticized by Erickson who, using the methodology of the paper of Stene et al (controlling by five year intervals) produced a strong paternal age effect in a data set which further appropriate analyses, by one year intervals, showed to be an artifact.[38] Similarly a strong effect was reported by Matsunaga et al for men 55 and over,[31] but Lamson et al showed the methodology used here was inappropriate.[39] Subsequent reanalysis of the data of Matsunaga et al by Cross et al (by one year intervals) revealed the trend originally reported to an increase in men 55 and over could still be found in the data set of Matsunaga et al but was weaker and no longer significant at the .05 level.** The remaining data sets on live births that report effects significant at the .05 level both indicate rather weak effects. [33,34] Interestingly, no significant effect or suggestive trend to an effect was found in British Columbia live births for 1952-63, although a positive effect was found for 1964-76.[33] (The data were divided in this way because it was believed ascertainment was almost complete in the second interval but not likely to be complete in the first.)

There are several hypotheses that may explain the discrepancies among the studies:

1. There are temporal, and/or geographic, and/or ethnic fluctuations in paternal age effects which have produced the variations among studies noted at present.
2. There is no paternal age effect and statistical fluctuation accounts for observations to date.
3. Methodological artifacts associated with incomplete sampling have obscured a paternal age effect in some studies that reported negative or non-significant trends.

---

*Stene J. Effect of advancing paternal age on the incidence of trisomy 21. Poster presentation at symposium on trisomy 21. Rapallo, Italy. November 8-10, 1979. In this study, out of 56 cases of Down syndrome diagnosed prenatally, 27 (48.2%) were in couples in which the fathers were 44 or over, whereas among the other 3485 cases, 833 (23.9%) were in couples with older fathers. Maternal age was controlled by one year intervals. See also addendumC

**Cross PK., Lamson SH, Hook EB. Unpublished analyses.

TABLE 5
Regression Derived Rates* of Down Syndrome per 1,000 Live Births at Selected Values of Maternal and Paternal Age: British Columbia, 1964-1976, Uncorrected for Underascertainment

Maternal Age	Paternal Age									All Paternal Ages**
	21	27	31	35	39	43	47	51	55	
21	0.5	0.6	0.6	0.6	0.6	0.7	0.7	0.7	0.7	0.6
23	0.6	0.6	0.6	0.7	0.7	0.7	0.8	0.8	0.8	0.6
25	0.6	0.7	0.7	0.7	0.8	0.8	0.8	0.9	0.9	0.7
27	0.7	0.8	0.8	0.8	0.9	0.9	0.9	1.0	1.0	0.8
29	0.8	0.9	0.9	0.9	1.0	1.0	1.0	1.1	1.1	0.9
31	0.9	0.9	1.0	1.0	1.1	1.1	1.2	1.2	1.3	1.0
33	1.2	1.3	1.3	1.4	1.5	1.5	1.6	1.6	1.7	1.4
35	2.0	2.1	2.2	2.3	2.4	2.5	2.6	2.7	2.8	2.4
37	3.3	3.5	3.7	3.8	4.0	4.1	4.3	4.5	4.7	4.0
39	5.5	5.8	6.1	6.3	6.6	6.8	7.1	7.4	7.7	6.7
41	9.1	9.6	10.0	10.4	10.9	11.3	11.8	12.2	12.7	11.3
43	15.0	15.9	16.5	17.2	17.9	18.6	19.3	20.1	20.9	18.9
45	24.6	26.1	27.1	28.2	29.3	30.4	31.7	32.9	34.2	31.5
47	40.1	42.5	44.1	45.9	47.6	49.5	51.4	53.4	55.5	52.0
49	64.8	68.5	71.1	73.8	76.6	79.5	82.4	85.5	88.7	84.7

*Ln $[y/(1-y)] = -7.1772 + 0.0533x_1 + 0.2528x_2 + 0.0100z$. The paternal age coefficient was fixed at 0.0100 (an intermediate value consistent with several data sets).

**Rates calculated from the equation in maternal age terms only. Ln $[y/(1-y)] = -6.8279 + 0.0608x_1 + 0.26162x_2$.

In the above equations: y = Rate of Down syndrome; z = Paternal age. x_1 & x_2 are functions of maternal age (x) as follows: $x_1 = x-32$ if $x \leqslant 32$ and 0 otherwise, and $x_2 = x-32$ if $x \geqslant 32$ and 0 otherwise.

TABLE 6

Regression Derived Rates of Down Syndrome per 1,000 Live Births at Selected Values of Maternal and Paternal Age: British Columbia, 1964-1976 Data, Increased by 20% to Adjust for Underascertainment*

Maternal Age	Paternal Age									All Paternal Ages**
	21	27	31	35	39	43	47	51	55	
21	0.6	0.7	0.7	0.7	0.8	0.8	0.8	0.8	0.9	0.7
23	0.7	0.7	0.8	0.8	0.8	0.9	0.9	0.9	1.0	0.8
25	0.8	0.8	0.9	0.9	0.9	1.0	1.0	1.1	1.1	0.8
27	0.9	0.9	1.0	1.0	1.0	1.1	1.1	1.2	1.2	1.0
29	1.0	1.0	1.1	1.1	1.2	1.2	1.2	1.3	1.4	1.1
31	1.1	1.1	1.2	1.2	1.3	1.3	1.4	1.4	1.5	1.2
33	1.5	1.5	1.6	1.7	1.7	1.8	1.9	2.0	2.0	1.7
35	2.4	2.6	2.7	2.8	2.9	3.0	3.1	3.2	3.4	2.8
37	4.0	4.2	4.4	4.6	4.8	5.0	5.2	5.4	5.6	4.8
39	6.6	7.0	7.3	7.6	7.9	8.2	8.5	8.9	9.3	8.1
41	10.9	11.6	12.0	12.5	13.0	13.6	14.1	14.7	15.3	13.5
43	18.0	19.1	19.8	20.6	21.4	22.3	23.2	24.1	25.1	22.7
45	29.5	31.3	32.5	33.8	35.1	36.5	38.0	39.5	41.0	37.8
47	48.1	51.0	53.0	55.0	57.2	59.4	61.7	64.1	66.5	62.4
49	77.7	82.2	85.3	88.6	91.9	95.4	98.9	102.6	106.4	101.6

*/**See footnotes to Table 5 for original regression equations used.

4. There is a weak paternal age effect in most if not all populations, but because of statistical fluctuation the trends are only significant in some data sets.

We believe there are too many suggestive indications of a paternal age effect to reject the existence of such an effect in all populations. We are, however, reluctant to postulate paternal age effects in some time periods and groups but not others without some biological grounds for believing such variation is plausible. Also, we find it difficult to postulate a methodological artifact that should obscure positive effects. Therefore, by exclusion we prefer the fourth hypothesis that postulates a slight increase in rate with paternal age which is obscured by statistical fluctuation in negative studies. To estimate such an effect, we note that a regression coefficient of about 1% per year for each year of paternal age at each fixed maternal age is consistent with the five data sets that have been subject to regression analyses.[33,36] Projections from this expectation appear in Table 5. These data have not been adjusted for presumptive underascertainment in the original study, so maternal age specific rates are lower than those in the Swedish study. We allow for this in Table 6. It is possible, of course, that there is a more complex paternal age effect than a simple first order exponential increase with paternal age. Many such can be suggested, but this is the simplest that appears consistent with all the data. (We suggest those doing analyses in the future use a similar regression analysis and report the confidence intervals for their regression coefficients for paternal age.) See also addendum C

What should genetic counselors do in the face of these or similar uncertainties until further discoveries are made? Discussion of this question could encompass an entire separate symposium! The rates given in Table 6 are subject to some uncertainty because of the assumptions used in the derivations. We would use them as guides in counseling as to risks associated with paternal age, but we recognize the limitations of the derived rates, and that not all readers may find this satisfactory.

RECURRENCE RISK

In the discussion of the next two sections we assume instances of parents with a 47,+21 mosaic line or Robertsonian translocation involving a 21 chromosome are excluded in the consideration of risk.

All the data on recurrence risk for 47,+21 Down syndrome after an affected live birth of which we are aware appear in Tables 7 and 8. These are based on calculations[12] from data presented by Stene.[40]

It may be noted that there are still several questions remaining, namely:

1. For a mother over 30 at risk for a second child with (47,+21) Down syndrome, what difference in risk is there according to the age at which

TABLE 7
Recurrence of Down Syndrome by Selected Maternal Ages*

	Maternal age at pregnancy at risk			
	<30	30-34	35-39	≥40
Proportion of affected cases:	3/211	1/145	0/165	1/112
	(1.4%)	(0.7%)	(0%)	(0.9%)

*Excludes those born to known 21 translocation carriers or 47,+21 mosaic parents. Calculations are as presented in reference 12 of analysis of data in reference 40.

TABLE 8
Recurrence of Down Syndrome*

Maternal age at birth of first case	Maternal age at risk		
	<30		≥30
<30	1.4% (0.3%–4.2%)	n.d.	⎫
			⎬ 0.4% (0.06%–1.7%)
≥30	–	n.d.	⎭

*n.d. = no data. Rates in parentheses are 95% confidence intervals.

she has had her first child with Down syndrome? (Available data are pooled over all maternal ages at birth of first affected child.)

2. What is the risk of a (47,+21) Down syndrome live birth if an earlier pregnancy has ended in:
   a. An abortus or stillbirth with 47,+21?
   b. A live birth with a cytogenetic abnormality other than 47,+21?
   c. An abortion or stillbirth with a cytogenetic abnormality other than 47,+21?
   d. An abortus or stillbirth of unknown genotype?

3. How do these risks vary with maternal age *at the time of the earlier pregnancy* and age at the time of risk?

With regard to question one, there are no data to my knowledge which allow an answer. Theoretically, it would appear that the younger the mother at the time of birth of the first affected child the higher her recurrence risk at any specific maternal age for a later pregnancy. The data for women 30 and over are pooled over all ages at time of birth of affected child. A conservative approach to counseling for recurrence at all maternal ages, suggested by Murphy and Chase, is to add about 1% to the age specific risk, whatever the age. This appears the most appropriate course at our present stage of knowledge.

With regard to 2a, the risk following a 47,+21 abortus or stillbirth, it appears most reasonable to counsel the woman as if the first pregnancy had ended in a viable live birth with 47,+21.

In situation 2b, a live birth with a cytogenetic abnormality other than 47,+21, there are to our knowledge no systematic data on this question. If an earlier pregnancy has ended in a *trisomy* other than 47,+21 one course is to counsel for Down syndrome risk as if it has ended in 47,+21. There are theoretical grounds for believing this is a conservative suggestion, and will give an upper bound for actual risk. Putative genes or environmental factors that induce non-disjunction of any chromosomes could induce heteroaneuploidy (occurrence of different chromosome abnormalities in the same sibship) or homoaneuploidy (recurrence of same abnormality).[41] Homoaneuploidy for 47,+21 could also be produced by putative factors specific for nondisjunction of the 21st chromosome, and also by (cryptic) parental mosaicism for 47,+21. Therefore, observed recurrence risks for 47,+21 may include components that do not apply to the risk of 47,+21 following another trisomy.

It could be argued that there are no firm data in *live births* clearly substantiating any increase in risk of 47,+21 following a previous pregnancy with chromosome abnormality other than 47,+21, but observation on spontaneous abortuses (see below) make it appear likely to us that such an increased risk may reasonably be inferred for live births.

With regard to 2c, the risk following an abortus with a non-47,+21 abnormality, the only data that come close to bearing on the risk of a subsequent 47,+21 live birth are those of Alberman et al.[42] They did not address the question directly because they had extensive information only on previous pregnancies of couples investigated. The rates of earlier Down syndrome in a couple with a history of a spontaneous abortus with any trisomy was 5/244 = 2.1%. Among sibs of abortuses with cytogenetic abnormalities other than trisomy the rate was 0/232 = 0%. Among sibs of abortuses with normal chromosomes the rate was 0/1042 = 0%. The rate was 0/1072 = 0% among siblings of live born controls. There are no data on subsequent pregnancies, but the data suggest strongly that a non-47,+21 trisomic spontaneous abortus is a risk factor for subsequent Down syndrome. Almost certainly this increased risk applies also following the occurrence of a stillbirth with such a cytogenetic outcome as well as an abortus.

Some further projections have been done by Lippman-Hand from pooled observations of studies of chromosome abnormalities in abortuses within sibships.[44] If 13.6% of pregnancies following trisomic loss result in abortion, 70.6% of these are trisomic, and 10% are 47,+21, then it is estimated that

0.96% of pregnancies subsequent to a trisomic abortion result in 47,+21. Using these figures, we calculate that if those lost as abortuses are about 2/3 of all 47,+21 conceptuses then about 0.4–0.5% of live births following trisomic abortions will have 47,+21. (0.7% is derived from a different assumption.[43]) This compares with a general risk of about 0.1% (over all ages). With regard to *any viable cytogenetic abnormality* following a trisomic abortus, the projection is 1%. Presumably the rate is lower for the occurrence of 47,+21 specifically, perhaps about 0.5%.

With regard to 2d, Lippman-Hand has also calculated from her assumptions, and the assumption that 25% of all abortuses are associated with (autosomal) trisomy, that the risk of a live birth having an (autosomal) trisomy following any abortion of *unknown genotype* is 0.43%.[43] (It may be calculated as 0.35% associated with 47,+21 and 0.09% with other autosomal trisomics.) The implications of this are truly extraordinary because this is the average risk of a woman age 35 or so. It may in fact be the case that a history of spontaneous abortions of unspecified type is a significant risk factor for 47,+21 live births. Nevertheless, it is strongly emphasized that the calculations are based on very sparse data derived from the study of heterogeneous populations, and most important, have not been corrected for maternal age or fetal gestational stage at the time of abortion. Before any application of such figures given above can be made to counseling situations we believe one must have data on the variability in the proportions in the baseline data by maternal age and gestational stage.

None of the analyses on recurrence risk have considered the adjustment for the number of normal pregnancies that intervene between the affected pregnancy and the one at risk. Other things being equal, a woman with a greater number of intervening normal outcomes will have a lower risk than a woman with none, but we do not have data at present to quantify this differential in risk.

With regard to the risk of a woman who has had an abortus of unknown genotype subsequently having a live born with 47,+21 or other cytogenetic abnormality, other things being equal such a woman who smokes and/or drinks alcoholic beverages and/or has other cause of embryonic and fetal death *not* associated with chromosome abnormality such as incompetent cervix, will have a *lower* risk of a cytogenetic abnormality in a future viable pregnancy. (This, it may be emphasized, is not to encourage smoking or alcoholic beverage ingestion. These agents apparently increase differentially the probability of embryonic or fetal death of cytogenetically normal conceptuses, consequently lowering the proportion of cytogenetic abnormalities in the increased total number that are lost.)

General conclusions regarding recurrence risks discussed in this section are summarized in Table 9.

TABLE 9
General Suggestions for Counseling for 47,+21 Down Syndrome
Live Birth "Recurrence"

Previous Outcome	Counsel
47,+21 live birth	By adding about 1% to maternal age specific risk.
47,+21 embryonic or fetal death	As if 47,+21 live birth occurred.
Live birth or fetal death with other 47, trisomy	Maximum risk as if 47, +21 live-birth occurred.
Live birth or fetal death with non-trisomic chromosome abnormality	? As if normal outcome has occurred (barring parental trans-location or mosaicism).
Spontaneous abortion of unknown genotype	? Intermediate risk between that associated with normal and tri-somic abortus.

RISK AFTER REMARRIAGE

Suppose one or both members of a couple who have had a 47,+21 Down syndrome child remarry. What is the risk for the new couple, particularly if one has no knowledge of the origin of the extra chromosome in the first child? There are to our knowledge no data assembled on this question, but it is not a rare occurrence. Without data, it is difficult to attempt an answer without making some strong assumptions and several weaker ones.

The strong assumptions are that the increased recurrence risk of a Down syndrome live birth to any couple is dependent only on factors in the parent, irrespective of sex, in whom the extra chromosome originated, that this risk travels to the new marriage, and that there is no interaction between parents in risk factors. The weaker assumptions are similar to those used in the more conventional counseling situations and apply to use of recurrence rates as recurrence risks.

An extensive discussion estimating risks on the basis of these assumptions has appeared elsewhere.[44]* If parental origin is known, it is suggested one counsel for increased risk only for that couple which includes the parent of origin of the original couple, whether father or mother. If parental origin is

*The discussion of paternal age in this reference is out of date, however.

unknown, assume about a 20–25% likelihood of paternal origin and 75–80% likelihood of maternal origin and apportion the excess risk accordingly. (The assumption that recurrence risk is independent of sex of parent of origin has been challenged.[45] To meet this objection models allowing for the extreme alternative possibilities that (a) it is only cases of *paternal* origin that contribute to the observed recurrence risk, and (2) it is only cases of *maternal* origin that contribute to the observed recurrence risk, have been considered. These models make opposite predictions that put limits on the true recurrence risk for any specific couple.[44])

While these suggestions appear reasonable, each reader independently must decide how plausible they are and whether there are other preferable alternatives in counseling, other than saying no risk at all can be cited.

ACKNOWLEDGMENTS

We thank Dina Schreinemachers, Lora Gersch, Robert Dorn, Linda Gulotty, and Scott Lamson for assistance over the past years in analysis of data reviewed here.

REFERENCES

1. Hook EB, Chambers GC. Estimated rates of Down syndrome in live births by one year maternal age intervals for mothers aged 20–49 in a New York study: Implications of the "risk" figures for genetic counseling and cost-benefit analysis of prenatal diagnosis programs. *In*: Bergsma D, Lowry RB, Trimble BK, Feingold M, eds. Numerical taxonomy of birth defects and polygenic disorders. Birth Defects Orig. Art. Ser. 13 (3a). New York: Alan R. Liss, Inc., 1977: 123-141.

2. Hook EB, Fabia JJ. Frequency of Down syndrome by single year maternal age interval: Results of a Massachusetts study. *Teratology* 1979; 17: 223-28.

3. Hook EB, Lindsjo A. Down syndrome in live births by single year maternal age interval in a Swedish study: Comparison with results from a New York study. *Am J Hum Genet* 1978; 30: 19-27.

4. Lamson SH, Hook EB. A simple function for maternal age specific rates of Down syndrome in the 20–49 age interval and its biological implications. *Am J Hum Genet* 1980; 32: 743-53.

5. Lamson SH, Hook EB. Comparison of mathematical models for the maternal age dependence of Down syndrome rates. *Hum Genet.* (in press)

6. Huether CA, Gummere GR, Hook EB, *et al*. Down syndrome: Percentage reporting on birth certificates and single year maternal age risk rates for Ohio 1970-79; comparison with Upstate New York data. *Amer J Pub Health*. 1981; 71: 1367-72.

7. Evans JA, Hunter AGW, Hamerton JL. Down syndrome and recent demographic trends in Manitoba. *J Med Genet* 1978; 15: 43-47.

8. Hook EB, Cross PK. Temporal changes in live birth prevalence of Down syndrome in New York State. *J Med Genet* 1981; 18: 29-30.

9. Hook EB, Cross PK. Surveillance of human populations for germinal cytogenetic mutations. *In*: Sugimura T, Kondo S, Takebe H, eds. Environmental mutagens and carcinogens. Tokyo/New York:University of Tokyo/ Alan R. Liss, Inc. (in press)

10. Hansson A, Mikkelsen M. The origin of the extra chromosome in Down syndrome. *Cytogenet Cell Genet* 1978; 20: 194-203.

11. Morris JN. Uses of epidemiology. 3rd ed. Edinburgh:Churchill Livingstone. 1975: 1-318. (See especially p. 98)

12. Hook EB. Down syndrome: Frequency in human populations and factors pertinent to variation in rates. *In*: de la Cruz F, Gerald PS, eds. Trisomy 21 (Down syndrome): Research perspective. Baltimore:University Park Press. 1981: 3-67.

13. Hook EB, Harlap S. Differences in maternal age-specific rates of Down syndrome between Jews of European origin and North African or Asian origin. *Teratology* 1979; 20: 243-48.

14. Hook EB, Porter IH. Human population cytogenetics — Comments on racial differences in frequency of chromosome abnormalities, putative clustering of Down syndrome, and radiation studies. *In*: Hook EB, Porter IH, eds. Population cytogenetics: Studies in humans. New York:Academic Press. 1977: 353-365.

15. Piper MC. The prevention of Down syndrome. Ph.D. Thesis. McGill University. 1979.

16. Fryers T, Mackay RI. Down syndrome: Prevalence at birth, mortality and survival. A 17 year study. *Early Hum Develop* 1979; 3: 29-41.

17. Mikkelsen M, Fischer G, Stene J, Stene E, Petersen E. Incidence study of Down syndrome in Copenhagen, 1960-71: With chromosome investigation. *Ann Hum Genet* 1976; 40: 177-82.

18. Griffiths AJF, Lowry RB, Renwick DHG. Down syndrome and maternal age in British Columbia, 1972-75. *Environ Health Persp* 1979; 31: 9-11.

19. Sutherland GR, Clisby SR, Bloor G, Carter RF. Down syndrome in South Australia. *Med J Austral* 1979: 2: 58-61.

20. Hook EB. Spontaneous deaths of fetuses with chromosomal abnormalities diagnosed prenatally. *New Eng J Med* 1978b; 299: 1036-38.

21. Polani PE, Alberman E, Berry AC, Blunt S, Singer JD. Chromosome abnormalities and maternal age. *Lancet* 1976; 2: 516-517.

22. Trimble BK, Baird PA. Maternal age and Down syndrome: Age specific incidence rates by single year intervals. *Amer J Med Genet* 1978; 2: 1-5.

23. Hook EB. The ratio of *de novo* unbalanced translocation to 47, trisomy 21 Down syndrome. A new method for human mutation surveillance and an apparent recent change in mutation rate resulting in human interchange trisomies in one jurisdiction. *Mutation Res* 1978; 52: 427-39.

24. Hook EB. Monitoring human mutations and consideration of a dilemma posed by an apparent increase in one type of mutation rate. *In*: Morton NE, Chung CS, eds. Genetic epidemiology. New York:Academic Press, 1978: 483-528.

25. Hook EB, Albright SG. Mutation rates for unbalanced Robertsonian translocations associated with Down syndrome in New York State live births 1968-77 and evidence for a temporal change. *Amer J Hum Genet* 1981; 33: 443-54.

26. Penrose LS. The relative effects of paternal and maternal age in mongolism. *J Genet* 1933; 27: 219-24.

27. Jenkins RL. Etiology of mongolism. *Amer J Dis Childh 1933; 44: 506.*

28. Mantel N, Stark ER. Paternal age in Down syndrome. *Am J Ment Defic* 1966; 71: 1025.

29. Carothers AD, Collyer S, DeMey R, Frackiewicz A. Parental age and birth order in the aetiology of some sex chromosome aneuploidies. *Ann Hum Genet* 1978; 41: 277-87.

30. Stene J, Fischer G, Stene E, Mikkelsen M, Petersen E. Paternal age effect in Down syndrome. *Ann Hum Genet* 1977; 40: 299-306.

31. Matsunaga E, Tonomura A, Oishi H, Kikuchi Y. Reexamination of paternal age effect in Down syndrome. *Hum Genet* 1978; 40: 259-68.

32. Stene J, Stene E, Stengel-Rutkowski S, Murken J-D. Paternal age and incidence of chromosomal aberrations: Prenatal diagnosis data (DFG) *In*: 6th Int Congr Human Genetics, Jerusalem-Abstracts, 1981: 49.

33. Hook EB, Cross PK, Lamson SH, Regal RR, Baird PA, Uh SH. Paternal age and Down syndrome in British Columbia. *Amer J Hum Genet* 1981; 33: 123-28.

34. Erickson JD, Bjerkedal T. Down syndrome associated with father's age in Norway. *J Med Genet* 1981; 18: 22-8.

35. Erickson JD. Paternal age and Down syndrome. *Am J Hum Genet* 1979; 31: 489-97.

36. Regal RR, Cross PK, Lamson SH, Hook EB. A search for evidence for a paternal age effect independent of a maternal age effect in birth certificate reports on Down syndrome in New York State. *Am J Epidemiol* 1980; 112: 650-55.

37. Liss SM, Hook EB. Search for paternal age effect independent of maternal age effect for cytogenetic abnormalities other than 47,+21 (sic) in New York State Chromosome Registry data. *Amer J Hum Genet* 1980; 32: 146A.

38. Erickson JD. Down syndrome, paternal age, maternal age and birth order. *Ann Hum Genet* 1978; 41: 289-98.

39. Lamson SH, Cross PK, Hook EB, Regal RR. On the inadequacy of analyzing the paternal age effect on Down syndrome rates using quinquennial data. *Hum Genet* 1980; 55: 49-51.

40. Stene J. Detection of higher recurrence risk for age-dependent chromosome abnormalities with an application to trisomy G (Down syndrome). *Hum Hered* 1970; 20: 112-22.

41. Hecht F. The non-randomness of human chromosome abnormalities. *In:* Hook EB, Porter IH, eds. Population Cytogenetics: Studies in humans. New York:Academic Press, 1977: 237-50.

42. Alberman E, Elliot M, Creasy M, Dhadial R. Previous reproductive history in mothers presenting with spontaneous abortions. *Br J Obstet Gynaecol* 1975; 82: 366-73.

43. Lippman-Hand A. Genetic counseling and human reproductive loss. *In:* Porter IH, Hook EB, eds. Human embryonic and fetal death. New York: Academic Press, 1979: 299-314.

44. Hook EB. Genetic counseling dilemmas: Down syndrome, paternal age, and risk after remarriage. *Am J Med Genet* 1980; 5: 145-55.

45. Karp LE. Editorial comments. See reference 44.

46. Stene J, Stene E, Stengel-Rutkowski S, Murken J-D, Paternal age and Down syndrome. *Human Genetics* 1981; 59: 119-24.

ADDENDA

Since this paper was written, several discoveries have been made which allow refinement, correction and interpretation of the data presented.

ADDENDUM A

Regarding Table 1, recently available data on the natural history of cytogenetically abnormal fetuses whose mothers do not select abortion provide an estimate of spontaneous fetal death rate of 30% (with 95% confidence interval of 19% to 42%) (EB Hook, submitted for publication). Therefore, the best value to use for adjustment of the amniocentesis rate in deriving an expected livebirth rate is 70%, not 80% as in the last column of Table 1.

ADDENDUM B

Regarding Table 2, recently we have discovered that not only is there a problem of completeness of birth certificates but also one of accuracy. We find about 5% to 10% of cases recently reported as Down syndrome on Upstate New York birth certificates are not affected. Some cases have other cytogenetic abnormalities such as 47, +18, some are just diagnostic errors. We do not have at present, evidence that this inaccuracy varies significantly by maternal age. Also we do not have any direct evidence that this has occurred in the Ohio study but we suspect strongly that it has. Our observations are on the recent years. We do not have evidence as yet that such inaccuracies occurred in 1963 to 1974 in Upstate New York but also suspect strongly they did. Such inaccurate reporting would imply that the rates in Table 2 tend to be slight overestimates, with the exception of the rates cited from the very first study.[1] For the latter, a tendency to an overestimate because of false "positives" may have been countered by a tendency to an underestimate because a completeness factor of 0.375 was used and not the lower value of 0.336 which was later found to be more appropriate. (See elsewhere in this paper.) We do not believe it likely that this problem of accuracy of certificates accounts for the apparent temporal change in our data for 1970-74 compared to 1965-69,[8] but it is very difficult at this point to investigate this possibility now.

ADDENDUM C

With regard to Tables 5 and 6 and the data on paternal age, these rates are acceptable, assuming the 1% increase with paternal age is correct. We have recently discovered however, that even analyses by one year interval may tend to produce artifactual paternal age effects, because of (a) the strong association of maternal age and paternal age, (b) the rise of the rate of Down syndrome with maternal age and (c) the fact that there is a very strong decline in fertility with maternal age. The average age at livebirth of women of truncated age 35 with a normal child is 35.4 and of a Down syndrome child is close to 35.6. For this reason alone, the expected value of case-control difference in paternal ages for livebirths born to women age 35 will be about +0.2 years, *i.e.* greater for the Down syndrome cases. Thus artifact and statistical fluctuation *may* be sufficient to account for all the observed trends in paternal age effects in the data reviewed. For this reason we suggest *assuming there are not temporal and geographical differences in putative paternal age effects* that the rates presented in Table 6 be regarded as guides to plausible *maximum* paternal age effects in genetic counseling.

It should be noted that recently Stene *et al.* [46] have published analyses of amniocentesis data claiming a very strong paternal age effect not only for trisomy 21 but for all chromosome aberrations, and that moreover, for a father age 41 or over, *irrespective* of the age of the mother, "prenatal diagnosis may be carried out. . .because we have found a strong paternal age effect from this age on."[46] Stene *et al.* had data on 60 Down syndrome cases in a total of 5014 studied. In data from the New York State Chromosome Registry however, on about twice as many cases, we find, using methods similar to those of Stene *et al.* as well as other methods no statistically significant evidence for any paternal age effects. (EB Hook, PK Cross, to be submitted.) Depending how the data are examined at best there are some suggestive trends but these may perhaps be explained by the artifact noted above. The results of Stene *et al.* may thus result from a combination of statistical fluctuation (chance) and statistical artifact. This does not necessarily mean that there is no paternal age effect upon Down syndrome, but only that on the basis of the cumulative weight of the evidence to date, such an effect does not appear established. Alternatively such effects may only be of significant biological importance in restricted geographical areas, (such as those that Stene *et al.* investigate).

We suggest that in the future those claiming statistically significant evidence for paternal age effects for Down syndrome examine to what extent the artifact we have cited could produce the effect or eliminate it by analyzing data by parental age in months or days at the time of conception.

APPENDIX: TABLE 7

Proportion of Mutant D/21 and G/21 Interchange Trisomies Among Down Syndrome Live Births in New York State 1968-80 Reported to New York State Chromosome Registry*

Year	Maternal Age	Mutant Frequency			Reported No. of DS** (lower)	Percent Mutants			Reported No. of DS† (upper)	Percent Mutants		
		D/21	G/21	All		D/21	G/21	All		D/21	G/21	All
1968	<30	2 (3.7)	1	3 (4.7)	51.1	3.9 (7.2)	2.0	5.9 (9.2)	55.1	3.6 (6.7)	1.8	5.4 (8.5)
	≥30	1 (1.3)	1	2 (2.3)	59.9	1.7 (2.2)	1.7	3.3 (3.8)	62.9	1.6 (2.1)	1.6	3.2 (3.7)
	All††	3 (5)	2	5 (7)	111.0	2.7 (4.5)	1.8	4.5 (6.3)	118.0	2.5 (4.2)	1.7	4.2 (5.9)
1969	<30	3	2	5	88.2	3.4	2.3	5.7	102.6	2.9	1.9	4.9
	≥30	0	1	1	85.8	0.0	1.2	1.2	98.4	0.0	1.0	1.0
	All	3	3	6	174.0	1.7	1.7	3.4	201.0	1.5	1.5	3.0
1970	<30	1.7	1 (2)	2.7 (3.7)	87.0	2.0	1.1 (2.3)	3.1 (4.3)	93.2	1.8	1.1 (2.1)	2.9 (4.0)
	≥30	1.3	0	1.3	103.0	1.3	0.0	1.3	108.8	1.2	0.0	1.2
	All††	3	1 (2)	4 (5)	190.0	1.6	0.5 (1.1)	2.1 (2.6)	202.0	1.5	0.5 (1.0)	2.0 (2.5)
1971	<30	2.7	2	4.7	99.8	2.7	2.0	4.7	106.3	2.5	1.9	4.4
	≥30	0.3	1	1.3	114.2	0.3	0.9	1.1	121.7	0.2	0.8	1.1
	All††	3	3	6	214.0	1.4	1.4	2.8	228.0	1.3	1.3	2.6

Year	Age									
1972	<30	0	2 (3)	109.0	0.0	1.8 (2.8)	112.9	0.0	1.8 (2.7)	1.8 (2.7)
	≥30	2	1	89.0	2.2	3.4	92.1	2.2	1.1	3.3
	All	2	5 (6)	198.0	1.0	2.5 (3.0)	205.0	1.0	1.5 (2.0)	2.4 (2.9)
1973	<30	4	8 (10)	103.3	3.9 (4.8)	7.7 (9.7)	107.9	3.7 (4.6)	3.7 (4.6)	7.4 (9.3)
	≥30	5	2	89.7	1.1	2.2	95.1	1.1	1.0	2.1
	All	5	10 (12)	193.0	2.6 (3.1)	5.2 (6.2)	203.0	2.5 (3.0)	2.5 (3.0)	4.9 (5.9)
1974	<30	5	10 (11)	114.6	4.4 (5.2)	8.7 (9.6)	121.6	4.1 (4.9)	4.1	8.2 (9.0)
	≥30	5	1	93.4	1.1	2.1	99.4	1.0	1.0	2.0
	All	6	12 (13)	208.0	2.9 (3.4)	5.8 (6.2)	221.0	2.7 (3.2)	2.7	5.4 (5.9)
1975	<30	4	9 (10)	110.0	3.6 (4.5)	8.2 (9.1)	116.1	3.4 (4.3)	4.3	7.8 (8.6)
	≥30	5	2	142.0	0.0	1.4	149.9	0.0	1.3	1.3
	All	4	11 (12)	252.0	1.6 (2.0)	4.4 (4.8)	266.0	1.5 (1.9)	2.6	4.1 (4.5)
1976	<30	1	3	101.3	1.0 (3.0)	4.6 (7.6)	103.6	1.0 (2.9)	3.6 (4.5)	4.5 (7.4)
	≥30	2	0	114.7	0.0	0.3	118.4	0.0	0.3	0.3
	All††	1	3	216.0	0.5 (1.4)	2.3 (3.7)	222.0	0.5 (1.4)	1.8 (2.3)	2.3 (3.6)

APPENDIX: TABLE 7, continued

Proportion of Mutant D/21 and G/21 Interchange Trisomies Among Down Syndrome Live Births in New York State 1968-80 Reported to New York State Chromosome Registry*

Year	Maternal Age	Mutant Frequency D/21	G/21	All	Reported No. of DS** (lower)	Percent Mutants D/21	G/21	All	Reported No. of DS† (upper)	Percent Mutants D/21	G/21	All
1977	<30	3	5 (6)	8 (9)	113.8	2.6	4.4 (5.3)	7.0 (7.9)	115.9	2.6	4.3 (5.2)	6.9 (7.8)
	≥30	0	2 (3)	2 (3)	88.2	0.0	2.3 (3.4)	2.3 (3.4)	91.1	0.0	2.2 (3.3)	2.2 (3.3)
	All	3	7 (9)	10 (12)	202.0	1.5	3.5 (4.5)	5.0 (5.9)	207.0	1.4	3.4 (4.3)	4.8 (5.8)
1978	<30	1 (2)	0	1 (2)	97.7	1.0 (2.0)	0.0	1.0 (2.0)	99.6	1.0 (2.0)	0.0	1.0 (2.0)
	≥30	3	1	4	106.3	2.8	0.9	3.8	108.4	2.8	0.9	3.7
	All	4 (5)	1	5 (6)	204.0	2.0 (2.4)	0.5	2.4 (2.9)	208.0	1.9 (2.4)	0.5	2.4 (2.9)
1979	<30	0 (5)	0	0 (5)	101.8	0.0 (4.9)	0.0	0.0 (4.9)	102.3	0.0 (4.9)	0.0	0.0 (4.9)
	≥30	2 (4)	0	2 (4)	85.2	2.3 (4.7)	0.0	2.3 (4.7)	85.7	2.3 (4.7)	0.0	2.3 (4.7)
	All	2 (9)	0	2 (9)	187.0	1.1 (4.8)	0.0	1.1 (4.8)	188.0	1.1 (4.8)	0.0	1.1 (4.8)
1980	<30	2	6	8	108.9	1.8	5.5	7.3	111.8	1.8	5.4	7.2
	≥30	0	3	3	81.1	0.0	3.7	3.7	83.2	0.0	3.6	3.6
	All	2	9	11	190.0	1.1	4.7	5.8	195.0	1.0	4.6	5.6

APPENDIX: TABLE 7, Footnotes

*NS = Not stated. DS = Down syndrome, all genotypes. Cases with maternal age not specified, of any genotype, were distributed into <30 and ≥30 categories according to the proportion of known cases in these categories. Numbers and rates in parentheses include presumed mutants and all those not known to be the result of familial translocation. Numbers and rates not in parentheses are on presumed mutants only. For categories in which there are no parenthetical entries, there were no cases in which familial origin could not be excluded.

**This coulumn includes only cases of DS of stated year of birth. Proportions derived using these figures are slight overestimates of the true proportions, except for 1971 in which they are slight underestimates.

†This column includes cases of known year of birth, and also cases of DS of unstated year of birth. The latter are assigned here to the year of report. The net result is a tendency to a slight underestimate of the true proportions except for 1971 in which they are slight overestimates.

††There was one D/21 case of unstated maternal age in each of the years 1968, 1970 and 1971. (The 1970 and 1971 cases are known mutants, the 1968 case is of unknown status.) For each such case 0.7 was assigned to the <30 groups and 0.3 to the ≥30 groups based on the distribution of mutants for whom maternal age was known. There was one G/21 known mutant of unstated maternal age born in 1976 treated similarly. The 1971 D/21 case was also the only interchange trisomy case reported of unstated year of birth (that could have been born in 1968-80). It is assigned to its year of report. For alternative treatment of cases of unstated maternal age and year of birth, see reference 25.

APPENDIX: TABLE 8
Proportion of Mutant D13 Interchange Trisomies Among Patau Syndrome
Live Births in New York State, 1968-80, Reported to the New York State
Chromosome Registry*

Year	Maternal Age	Mutant Frequency	Reported Patau Synd.	Percent Mutant
1968	<30	0	0	0.0
	≥30	0	1	0.0
	All	0	1	0.0
1969	<30	0 (.7)**	2	0.0 (35.0)
	≥30	0 (.3)	2	0.0 (15.0)
	All	0 (1)	4	0.0 (25.0)
1970	<30	0.7**	2.5	28.0
	≥30	0.3	2.5	12.0
	All	1	5.0	20.0
1971	<30	1	5	20.0
	≥30	0	5	0.0
	All	1	10	10.0
1972	<30	0 (1)	4.7	0.0 (21.3)
	≥30	1	2.3	43.5
	All	1 (2)	7.0	14.3 (28.6)
1973	<30	2	7.7	26.0
	≥30	0	1.3	0.0
	All	2	9.0	22.2
1974	<30	3.7**	10.5	35.2
	≥30	0.3	3.5	8.6
	All	4	14.0	28.6

APPENDIX: TABLE 8, continued

Year	Maternal Age	Mutant Frequency	Reported Patau Synd.	Percent Mutant
1975	<30	1	15.6	6.4
	≥30	2	2.4	83.8
	All	3	18.0	16.7
1976	<30	0	6.0	0.0
	≥30	1	6.0	16.7
	All	1	12.0	8.3
1977	<30	1.7**	11.4	14.9
	≥30	0.3	1.6	18.8
	All	2	13.0	15.4
1978	<30	2	11.4	17.5
	≥30	1	4.6	21.7
	All	3	16.0	18.8
1979	<30	2 (3)	9.6	20.8 (31.3)
	≥30	0	7.4	0.0
	All	2(3)	17.0	11.8(17.6)
1980	<30	0 (1)	8.4	0.0 (11.9)
	≥30	1	3.6	27.8
	All	1 (2)	12.0	8.3 (16.7)

*PS = Patau syndrome. Numbers and rates in parentheses are on known mutants and those in which a parental translocation has not been excluded. Numbers and rates not in parentheses are on known mutants only.

**Interchange trisomic cases of unstated maternal age occurred in 1969, 1970, 1974 and 1977. Based on the known distribution of maternal age for the other D/13 interchange trisomies, 0.7 of each such case was assigned to the <30 category, and 0.3 of each such case to the ≥30 category.

GENETIC COUNSELING FOR NORMAL PARENTS
WITH TWO OR MORE RETARDED CHILDREN:
A DIAGNOSTIC DILEMMA

Lawrence R. Shapiro
Patrick L. Wilmot
Murray D. Kuhr
Evelyn Lilienthal
Linda C. Higgs

When normal parents have two or more mentally retarded children, it is important to attempt to establish an accurate diagnosis, elucidate genetic mechanisms, and ultimately provide genetic counseling. Unfortunately, mental retardation in two or more siblings with two normal parents is a subject about which little is known, and often presents a diagnostic dilemma.

After evaluating a number of such families, it was our impression that we were frequently not able to make an accurate diagnosis or define genetic mechanisms. Two major prior studies of familial mental retardation were not helpful because families with mentally retarded parents as well as children were included.[1,2] The general recurrence risk for non-specific mental retardation among siblings after one index case has been calculated to be 3.7% increasing to 12% after two affected siblings.[3] While these estimates are helpful in approaching multiple sibling mental retardation from a genetic counseling standpoint, they do not take into account specific diagnoses or etiologies.

Because of the difficulties we encountered in attempting genetic counseling for normal parents with two or more mentally retarded children and the absence of helpful information, a study was undertaken to review our experience and develop an approach to the dilemma presented in diagnoses and genetic counseling.

The X-linked recessive form of mental retardation associated with a fragile X chromosome was reported near the conclusion of this study.[4,5] Consequently, most families with only affected males and other selected families were reevaluated for the presence of the Xq28 fragile site.

TABLE 1
Ascertainment of Families with
Multiple Sibling Mental Retardation

Families considered	70
Institutionalized	45/70 (64.3%)
Community	25/70 (35.7%)

TABLE 2
Investigations[†]

Pedigrees	70/70
Physical examination	68/70
Chromosome analysis	68/70
prior to banding	2/68
Thyroid studies	50/70
Amino acid studies	
blood	57/70
urine	55*/70
Dermatoglyphics	68/70

†Number of families investigated/
total number of families.
*Included mucopolysaccharide spot
test.

TABLE 3
Result by Diagnosis and/or Genetic Mechanisms*

Chromosome aberration excluding fra (X)		4/70 (5.8%)
All unbalanced translocations resulting from balanced carriers:		
	partial 9p+	(1/70)
	partial 5q+	(1/70)
	partial 2q+	(1/70)
	D/G Down syndrome	(1/70)
Autosomal recessive		11/70 (15.7%)
Autosomal dominant		1/70 (1.4%)
X-linked recessive without fra (X)		1/70 (1.4%)
X-linked recessive with fra (X)		10/70 (14.3%)
Unclassifiable		43/70 (61.4%)

*Number families diagnosed/total number of families.

DEFINITION

For the purpose of this study, multiple sibling mental retardation is defined as two or more mentally retarded children with normal parents.

SUBJECTS

A total of 76 families with 174 individuals were ascertained. Six families with 20 individuals were eliminated because of specific environmental factors such as child abuse, alcoholism, drug abuse, psychiatric disorders, *etc.* Alcoholism in one or both parents was viewed as a social/enviornmental factor rather than a possible teratogenetic influence because of the lack of a specific history of alcohol intake *during* pregnancy and no apparent clinical diagnosis of the fetal alcohol syndrome in the families evaluated.

Thus, 70 families with 154 individuals, of which 125 individuals were actually seen, were studied.

ASCERTAINMENT AND METHODS

Of the 70 families studied, 45 (64.3%) were residents in a large institution for the mentally retarded and 25 (35.7%) were ascertained from within the community (Table 1).

Investigations included pedigree analysis and/or record review, complete physical examination, chromosome analysis, thyroid studies, amino acid studies and dermatoglyphic analysis (Table 2). The two families who were not examined and who did not have chromosome analysis or dermatoglyphic analysis were known cases of phenylketonuria, the diagnosis being established prior to ascertainment.

Once the X-linked recessive form of mental retardation associated with a marker X chromosome was reported,[4,5] 20 of 25 families with no diagnosis and only affected males and five other selected families for which no diagnosis had been established were evaluated for the presence of an Xq28 fragile site.

RESULTS

As indicated in Table 3, chromosome aberrations occurred in 5.8% of the total cases. All were unbalanced translocations resulting from balanced translocation carrier parents. Autosomal recessive disorders were found in 15.7%, while autosomal dominant and X-linked recessive disorders without the fragile X chromosome occurred 1.4% each. An Xq28 fragile site was found in ten of the 70 families (14.3%). One of the families with the X-linked recessive form of mental retardation associated with the Xq28 fragile site was determined to have X-linked recessive inheritance on the basis of the pedigree prior to fragile X detection.

TABLE 4

Classification by Diagnosis and/or Genetic Mechanism

Autosomal Dominant:	
Myotonic dystrophy	1 family
Autosomal Recessive:	
a. Biochemical/metabolic	
PKU	3 families
"Degenerative" neurological disorder	2 families
Cretin	1 family
b. Other	
Consanguinity	3 families
Laurence–Moon–Biedel	1 family
Microcephaly, nephrosis, hiatal hernia syndrome	1 family

TABLE 5

Impact of Fragile X Chromosome*

	Prior to fra (X)	After fra (X)
Known diagnosis	18/70 (25.7%)	27/70 (38.6%)
Known diagnosis-institution	8/45 (17.8%)	15/45 (33.3%)
Known diagnosis-community	10/25 (40%)	12/25 (48%)
Families with affected males only*	6/25 (24%)	15/25 (60%)

*Number families known diagnosis/total number families.

Of the 70 families studied, 43 (61.4%) were not classifiable; no specific etiology or genetic mechanism could be determined. A classification of autosomal recessive inheritance is not possible unless a known condition is found or consanguinity exists; however, for genetic counseling purposes, the possibility must be considered. Table 4 lists the autosomal dominant and autosomal recessive conditions detected in this study.

IMPACT OF THE FRAGILE X CHROMOSOME

The importance and impact of the fragile X chromosome can be seen when the results of this study are compared *prior* to detection of the fragile X chromosome and after testing for the presence of the Xq28 fragile site. (Table 5) Prior to testing for the fragile X chromosome, a known diagnosis or genetic mechanism was determined in 18/70 families (25.7%). However, after testing for the Xq28 fragile site, the percentage of known diagnoses increased to 38.6%. The major increase in diagnoses appeared to occur for those families who were ascertained within the large institution for the mentally retarded, rather than those referred from within the community. This may represent an ascertainment bias.

For those families with only affected males who were evaluated prior to testing for the fragile X chromosome, a diagnosis or genetic mechanism was determined in 24%, however, after detection of the Xq28 fragile site, this percentage increased to *60%*. Unquestionably, the fragile X chromosome is extremely important in those families with only affected males or a predominance of affected males. No mentally retarded females with a fragile X chromosome[6] were detected in our study. This is probably due to the small size.

SUMMARY

For all families and especially those with only affected males, the ability to detect the Xq28 fragile site results in a marked increase of definite diagnoses and elucidation of genetic mechanisms. While the ability to detect the Xq28 fragile site has a major impact on the difficult clinical problem represented by multiple sibling mental retardation, the fact that 61.4% of the families in this study could not be classified with regard to etiology or genetic mechanism reflects the diagnostic dilemma posed by these families and the difficulties of genetic counseling.

ACKNOWLEDGMENTS

This paper supported in part by grants from the New York State Department of Health, Birth Defects Institute and the March of Dimes Birth Defects Foundation.

REFERENCES

1. Reed SC, Reed EW. Mental retardation, a family study. Philadelphia: Saunders, 1965.
2. Rett A. Clinical, genetic and psychological findings in 600 families with more than one mentally retarded child. *In*: Mittler P, ed. Research to practice in mental retardation — Biomedical aspects. Vol. III Baltimore: University Park Press, 1977.
3. Herbst DS, Baird PA. Recurrence risks for nonspecific mental retardation. *Amer J Hum Genet* 1981; 33: 80A.
4. Howard-Peebles PN, Stoddard GR, Mims MG. Familial X-linked mental retardation, verbal disability, and marker X chromosomes. *Amer J Hum Genet* 1979; 31: 214.
5. Turner G, Daniel A, Frost M. X-linked mental retardation, macroorchidism and the Xq27 fragile site. *J Pediatr* 1980; 96: 837.
6. Turner G, Brookwell R, Daniel A, Selikowitz M, Zilibowitz M. Heterozygous expression of X-linked mental retardation and X chromosome marker fra (X) (q27). *N Engl J Med* 1980; 303: 662.

THE LARGE-FOR-GESTATIONAL-AGE (LGA) INFANT
IN DYSMORPHIC PERSPECTIVE

M. Michael Cohen, Jr.

In genetics and pediatrics a great deal of attention has been focused on intrauterine growth retardation which may occur alone or with a great many anomalies, making up various syndromes. Relatively little attention has been paid, however, to overgrowth and its attendant syndromes. There are three reasons for this: First, overgrowth syndromes are much less common than syndromes with growth deficiency. Second, of the few known overgrowth syndromes, most of them have been poorly delineated until relatively recently. Finally, some remain completely undelineated.

There are a number of different ways to classify overgrowth.[1] Growth excess may be categorized by its time of onset (either prenatal or postnatal), and further categorized as either primary (representing intrinsic cellular hyperplasia) or secondary (representing overgrowth caused by humorally mediated factors outside the skeletal system).

Normal variants of growth include most instances of infant macrosomia above 4,000 grams which account for approximately 5% of all newborns. (Figure 1). Various factors implicated in large birth weight infants have been reviewed by Cohen.[2] Normal variants also include the polygenically determined upper extreme of normal growth: familial tall stature and familial rapid maturation.[1] In pathologic overgrowth, most prenatal onset growth excess is of the primary type, characterized by cellular hyperplasia as in the Beckwith-Wiedemann syndrome. There are a few examples of primary growth excess of postnatal onset such as the XYY syndrome. Postnatal onset growth excess, however, is most commonly secondary, involving humorally mediated factors such as the early production of estrogens, androgens, or both in sexual precocity. Secondary growth excess of prenatal onset is distinctly uncommon in terms of number of such syndromes, diabetic macrosomia serving as an example.

This paper considers only overgrowth of prenatal onset. Such overgrowth may result from (a) an increased number of cells, (b) hypertrophy, (c) an increase in the interstitium (most commonly excessive fluid as in anasarca), or (d) some combination of the above. In the overgrowth syndromes discussed in this paper, excessive cellular proliferation predominates and can be demon-

CLINICAL GENETICS: PROBLEMS IN
DIAGNOSIS AND COUNSELING

153

TABLE 1
Relationship of Overgrowth Syndromes to Neoplasia

Condition	Reported neoplasms	No. reported patients	Type of Overgrowth
Diabetic macrosomia	No	Many	Secondary
Infant giants	No	Few	Secondary
Beckwith-Wiedemann	Yes	Many	Primary
Hemihypertrophy	Yes	Many	Primary
Sotos	Yes	Many	Primary
Nevo	No	Few	Primary
Ruvalcaba-Myhre	Yes	Few	Primary
Weaver	?*	Few	Primary
Elejalde**	No	Few	Primary

*One presumed case possibly may be associated with a neuroblastoma. Less than a dozen cases are known and not all of these have been reported to date.

**None of the three known cases were compatible with life.

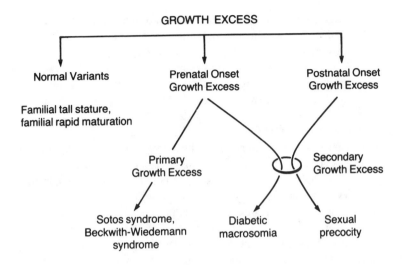

Fig. 1. Growth excess.

strated or inferred to have occurred. These syndromes tend to share several characteristics. First, overgrowth is commonly present at birth and persists into postnatal life. Second, weight is as important as length. Third, most of these overgrowth syndromes are associated with various anomalies. Fourth, mental deficiency is often a feature. Finally, some of these syndromes are associated with neoplasia.[2]

The association of overgrowth syndromes with neoplasia (Table 1) is not surprising since both processes represent an increase in the number of cell divisions.[2] Children who develop Wilms' tumor tend to have higher birthweights.[3] Osteosarcoma tends to arise in bones that grow rapidly and produce taller individuals.[4] Larger breeds of dogs are more susceptible to osteosarcoma than smaller breeds.[5] Those syndromes not known to be associated with neoplasia can be explained in one of two ways. First, with some overgrowth syndromes, too few cases have been reported to establish or exclude relationships to neoplasia with certainty. Second, a condition like diabetic macrosomia is common, yet is not known to be associated with an increase in neoplasia. However, overgrowth in this condition is secondary (humorally mediated) in contrast to most prenatal onset overgrowth syndromes whose growth is primary in type (intrinsic cellular hyperplasia).

The following is a review of known overgrowth syndromes of prenatal onset.[2]

DIABETIC MACROSOMIA

The frequency of macrosomic infants born to diabetic mothers varies from 7% to 19%, depending upon the particular study and upon the birthweight above which macrosomia is arbitrarily defined. The mean birthweight of infants born to diabetic mothers is 500 grams heavier than the birthweight of infants born to nondiabetic mothers.[6] Of 352 infants of diabetic mothers in the Cleveland study, 85% had birthweights above the norm for gestational age.[7] The likelihood of finding maternal diabetes mellitus increases with the degree of fetal macrosomia.[6] Certainly, gestational diabetics are prone to have macrosomic infants. Long-term follow-up indicates that a significant proportion of these mothers go on to develop symptomatic diabetes mellitus.

The pathogenesis of diabetic macrosomia is based on the chain of events that follows maternal hyperglycemia. The resultant fetal pancreatic hyperfunction leads to macrosomia since insulin is growth-promoting. Although insulin is strongly lipotropic and some authors have claimed that diabetic macrosomia is based primarily on excessive adipose tissue, nonadipose body constituents are increased in macrosomic infants of diabetic mothers.[8] Such infants also have an increased number of cell nuclei in various organs, suggesting increased protein synthesis most easily attributable to anabolic effects of insulin other than lipogenesis.

The size of infants born to a diabetic mother may vary with each pregnancy. Such differences are probably related to the balance between fetal pancreatic hyperfunction on the one hand, and vascular insufficiency with placental dysfunction on the other. Hyperinsulinism tends to result in macrosomia, while vascular insufficiency with placental dysfunction tends to result in microsomia. Adolescent obesity is more likely to occur in LGA infants of diabetic mothers than in LGA infants of normal mothers,[9] which suggests that macrosomia in infants of diabetic mothers may be a predisposing factor for later obesity.

INFANT GIANTS

Infant giants represent a rare disorder in which gross macrosomia occurs without associated malformations. Overgrowth is based on prenatal hyperinsulinism that persists into postnatal life. There is never a history of maternal diabetes and maternal glucose tolerance tests are always normal. Affected infants have severe intractable hypoglycemia that is refractory to various types of medical therapy, and total pancreatectomy is required for control. Birthweights vary from approximately 3,800 grams to well over 5,000 grams. Onset of symptoms varies from birth to 48 hours of age, with the majority occurring between 17 and 40 hours. Of the nine reported cases before 1966, death occurred between 35 hours and 4½ years of age. Subsequently, survivors have been reported, but they are usually moderately to severely retarded. Only 14

TABLE 2

Estimated Frequencies of Clinical and Laboratory Findings
in the Beckwith-Wiedemann Syndrome*

	%
Growth/Skeletal	
Increased birthweight	39
Postnatal gigantism	33
Accelerated osseous maturation	21
Asymmetry	13
Skeletal anomalies	14
Performance	
Seizures, apnea, cyanosis	22
Mental deficiency	12
Craniofacial	
Macroglossia	82
Ear lobe grooves	38
Flame nevus	32
Craniofacial dysmorphism**	39
Mild microcephaly	14
Abdominal/Genitourinary	
Omphalocele, umbilical hernia	75
Gastrointestinal anomalies	13
Hepatomegaly	32
Splenomegaly	14
Nephromegaly	23
Genitourinary anomalies	24
Cardiac anomalies	16
Diaphragmatic anomalies	7
Inguinal hernia	6
Laboratory Findings	
Hypoglycemia	30
Polycythemia	20
Hypocalcemia	5
Hypercholesterolemia, hyperlipidemia	2

*Frequencies estimated from 174 cases from the literature tabulated by C. Sotelo-Avila et al. See Ref. 14. There is an obvious ascertainment bias in the frequencies of various findings from reported cases. On the one hand, the more severe cases are most likely to be reported. On the other hand, the presence or absence of various findings may be omitted from some reports.
**Includes maxillary hypoplasia, prominent occiput, flat nasal bridge, highly arched palate, frontal ridge, downslanting palpebral fissures, etc.

cases had been reported by 1976.[10] Of these, affected sibs of a consanguineous union were reported;[11] all other instances were sporadic, at least at the time of reporting. It is possible that the disorder is an autosomal recessive condition. Because of the small size of the human family, more than half of all autosomal recessive disorders occur sporadically. Futhermore, a genetic condition may sometimes appear to be almost exclusively sporadic during the early stages of its delineation. Seventy-five percent of infant giants have been girls. This may possibly reflect (a) small sample size, or (b) an early lethal effect on homozygous male embryos. Further reports of infant giant sibs in which both sexes are affected would strengthen the autosomal recessive hypothesis.

Although other familial cases of hypoglycemia are known,[10] they represent conditions that are different from infant giants. The only other disorder with both hypoglycemia and overgrowth of prenatal onset is the Beckwith-Wiedemann syndrome; some of these cases are known to be familial. A few examples of infant giants may represent the Beckwith-Wiedemann syndrome without associated malformations. Instances of the Beckwith-Wiedemann syndrome with overgrowth in the absence of malformations have been observed in which the visceral histologic lesions of the syndrome were florid. Thus, for some infant giants to represent the Beckwith-Wiedemann syndrome, the histologic lesions of the latter must be found in the former. This hypothesis can be confirmed or rejected by histologic studies of the viscera at future postmortem examinations.

BECKWITH–WIEDEMANN SYNDROME

In 1963, Beckwith reported three cases of a newly recognized syndrome consisting of macroglossia, omphalocele, cytomegaly of the adrenal cortex, hyperplasia of the gonadal interstitial cells, renal medullary dysplasia, and hyperplastic visceromegaly.* Subsequently, Beckwith[12] enlarged his series of patients, noting postnatal somatic gigantism, mild microcephaly, and severe hypoglycemia. In 1964, Wiedemann[13] independently reported the syndrome in three sibs and observed a further component, a dome-shaped defect of the diaphragm. Other important contributions have been made by many investigators, and are reviewed elsewhere.[2] Early diagnosis of this striking condition alerts the clinician to the dual threat of hypoglycemia and possible neoplasia[+].

Gigantism is not always present at birth. Growth may even be below normal for a few months, but somatic gigantism eventually results in most cases. The height and weight are often above the 90th percentile. Advanced

---

*Extreme cytomegaly of the adrenal fetal cortex, omphalocele, hyperplasia of the kidneys and pancreas, and Leydig-cell hyperplasia: Anther syndrome? Presented at the annual meeting of the Western Society for Pediatric Research, Los Angeles, Nov. 11, 1963.

+See Table 2.

bone age is usually present and, in some cases, there may be widening of the metaphyses and cortical thickening of the long bones. Hemihypertrophy has been a feature in approximately 13% of the cases.[14]

Of the over 200 reported cases of the Beckwith-Wiedemann syndrome, 20 patients had neoplasms: 17 were malignant and 5 benign. Wilms' tumor occurred most commonly, followed by adrenal cortical carcinoma. (Table 3) Thirty percent of these patients had hemihypertrophy.[14] The overall frequency with which tumors occur in the Beckwith-Wiedemann syndrome is unknown because of the reporting bias favoring cases in which neoplasms have developed. Most cases of the syndrome are probably free of neoplasia.

Cases of the Beckwith-Wiedemann syndrome tend to be sporadic, but familial instances have been reported on numerous occasions.[2,15] Based on familial occurrence, autosomal recessive, autosomal dominant, autosomal dominant-sex-dependent, and multifactorial modes of inheritance have been proposed. The numerous occurrences of affected sibs certainly suggest an autosomal recessive mode of inheritance. However, recessive inheritance is incompatible with many pedigrees including an affected mother and child, and several instances of affected sibships in the same family. All of these pedigrees are compatible with autosomal dominant inheritance with low-penetrance and variable expressivity; some individuals exhibit only minor anomalies such as ear lobe grooves. Herrmann and Opitz[16] used the concept of delayed mutation to explain cases of the Beckwith-Wiedemann syndrome in sporadic cases, in affected sibs, in consecutive generations, and in more distantly related family members. Multifactorial inheritance has been discussed by Gardner[17] and Berry et al.[18] Because several authors have noted the occurrence of diabetes mellitus in nonaffected family members of Beckwith-Wiedemann patients, it is possible that altered carbohydrate metabolism might be a precipitating maternal factor. Thus, the possible interaction of "at risk" genes in the fetus with metabolic factors in the mother suggests multifactorial control. The possibility should be kept in mind that the Beckwith-Wiedemann syndrome might be etiologically heterogeneous.

Prenatal diagnosis would be useful for the Beckwith-Wiedemann syndrome. Although omphalocele has been detected by alpha-fetoprotein, many patients with the Beckwith-Wiedemann syndrome do not have omphalocele. Shapiro (L.R. Shapiro, personal communication) has detected a large placenta prenatally by ultrasound in the Beckwith-Wiedemann syndrome. At present, it is not known if all or most Beckwith-Wiedemann fetuses have large placentas and further investigation in this area is necessary.

HEMIHYPERTROPHY

Although the term hemihypertrophy has been used conventionally and frequently in the medical literature, it is inappropriate since the condition so obviously refers to hemihyperplasia. The differences between common asym-

TABLE 3

Neoplasms Associated with Overgrowth Syndromes*

	Beckwith-Wiedemann syndrome**	Hemihypertrophy**	Sotos syndrome
Malignant	Nephroblastoma† Adrenal cortical carcinoma† Hepatoblastoma† Hepatocellular carcinoma Glioblastoma Rhabdomyosarcoma	Nephroblastoma † Adrenal cortical carcinoma † Hepatoblastoma† Adrenal neuroblastoma Pheochromocytoma Undifferentiated sarcoma	Nephroblastoma Hepatocellular carcinoma
Benign	Adrenal adenoma Carcinoid tumor Fibroadenoma Fibrous hamartoma Ganglioneuroma Lipoma Myxoma	Adrenal adenoma	Cavernous hemangioma Hairy pigmented nevus Osteochondroma Pleomorphic adenoma

*See Ref. 2.

**See Ref. 14, for discussion of the relationship between Beckwith-Wiedemann syndrome and hemihypertrophy.

†Most commonly observed neoplasms in descending order of frequency.

metry, hemihyperplasia, hemiatrophy, and preferential laterality have been discussed elsewhere.[19] In hemihypertrophy, the enlarged area may vary from a single digit, a single limb, or unilateral facial enlargement to involvement of half the body.[20] Hemihypertrophy may be segmental, unilateral, or crossed. In some cases, the defect is limited to a single system, *e.g.*, muscular, vascular, skeletal, or nervous system, but it may frequently involve multiple systems. Asymmetry is usually evident at birth and may become accentuated with age, especially at puberty. Occasionally, asymmetry has been stated not to be present at birth, but to develop later. However, such observations are valid only when measurements are taken at birth. In total hemihypertrophy, the right side is more frequently involved and males are more frequently affected.

The bones have been found to be unilaterally enlarged, and increased bone age on the affected side has been reported. Other abnormalities of the skeletal system may include macrodactyly, syndactyly, polydactyly, compensatory scoliosis, and a variety of other skeletal defects. Thickened skin in the affected area, excessive secretion of sebaceous and sweat glands, vascular and pigmentary defects, hirsutism, hypertrichosis of the affected side, and abnormal nail growth have been reported. Central nervous system defects may include unilateral enlargement of a cerebral hemisphere, mental retardation, and seizures. Various accompanying anomalies have been noted elsewhere.[21]

Neoplasms are associated with hemihypertrophy (Table 3). Most common are Wilms' tumor, adrenal cortical carcinoma, and hepatoblastoma. Since these tumors have embryonal origin, study of their relationship to the teratogenic aspects of hemihypertrophy may lead to more specific information on oncogenic mechanisms. There is no relationship between laterality of hemihypertrophy and any of the solid tumors reported. When the oncogenic stimulus does not lateralize to the enlarged side, it is possible that such cases may represent "occult" crossed hemihypertrophy, affecting the internal organs of the contralateral side.

The etiology and pathogenesis are poorly understood. A tendency toward dizygotic twinning has been observed in some cases. Chromosomal anomalies, including diploid/triploid mosaicism, trisomy 18 mosaicism, and partial G and B monosomy with B/G translocation, and abnormally large chromosome number 3 have been reported.[22] Many other theories have been advanced to explain hemihypertrophy, including anatomic and functional vascular or lymphatic abnormalities, lesions of the central nervous system leading to altered neurotrophic action, endocrine abnormalities, asymmetric cell division and deviation of the twinning process, fusion of two eggs following fertilization leading to unequal regulative ability in the two halves, and mitochondrial damage to an overripened egg leading to overregeneration.[23] The range and variability of clinical abnormalities, together with the large number of sporadic cases, suggest etiologic heterogeneity. None of the proposed theories, some of which are quite fanciful, explains adequately all cases of hemihypertrophy.

Almost all cases appear to be sporadic. Familial instances, frequently incompletely documented, may represent other disorders, particularly neurofibromatosis.

The Beckwith-Wiedemann syndrome overlaps with hemihypertrophy; the two conditions cannot be totally separated. The Beckwith-Wiedemann syndrome and hemihypertrophy are probably at either end of the same spectrum, intermediate forms being the connecting links.[14] Some classic cases of the Beckwith-Wiedemann syndrome are associated with hemihypertrophy. Furthermore, both conditions share many features, such as medullary sponge kidney, anomalies of the urinary system, cryptorchidism, and malignant tumors including Wilms' tumor, adrenal cortical carcinoma, and hepatoblastoma. The tissue and organ hyperplasia of the Beckwith-Wiedemann syndrome may at times be generalized, at other times be localized to one side of the body, and at still other times be confined to certain organs and tissues.[15] It should be recognized, however, that hemihypertrophy is known to be etiologically heterogeneous. Thus, not all cases of hemihypertrophy are necessarily part of the Beckwith-Wiedemann spectrum. For example, chromosomal mosaicism has been reported with some cases of hemihypertrophy (*vide supra*).

SOTOS SYNDROME

Well over 100 cases of cerebral gigantism or Sotos syndrome had been reported by 1977.[24] Jaeken *et al*[25] reviewed 80 cases and tabulated the frequencies of the most commonly reported clinical findings (Table 4). The mean birthweight is 3,400 g(75th to 90th percentile) and mean birth length is 55.2 cm(97th percentile = 54.2 cm). Growth velocity is especially excessive during the first two or three years of life. Bone age is usually two or three years in advance of chronologic age. After three years of age, growth velocity proceeds at a more normal rate.

Most patients have had nonprogressive neurologic dysfunction, manifested by unusual clumsiness and dull intelligence. The mean I.Q. is 60. At times, behavior may be aggressive. Dilatation of the cerebral ventricles is common. Seizures, respiratory and feeding problems have been noted in some patients. Delay in walking until after 15 months of age and in the development of speech until after two and one half years are usual. Neoplasms are associated with Sotos syndrome[14,22] (Table 3).

A 14% incidence of glucose intolerance has been demonstrated in Sotos syndrome by numerous investigators. Nineteen percent of families include members with diabetes mellitus. Growth hormone studies have been normal except for one patient who had a paradoxical rise in growth hormone level in response to hyperglycemia, which suggests hypothalamic disregulation. Normal peripheral sensitivity to growth hormone is suggested by the demonstration of a normal rise in nonesterified fatty acid levels after the administration of growth hormone. Studies of somatomedin in Sotos syndrome have

TABLE 4
Clinical Findings in Sotos Syndrome

	%
Growth/Skeletal	
Gigantism	100
Accelerated osseous maturation	74
Large hands and feet	83
Performance	
Developmental retardation	83
Lack of fine motor control	67
Neonatal adaptation and/or feeding difficulties	44
Craniofacial	
Dolichocephaly	84
Prominent forehead	96
Ocular hypertelorism	91
Pointed chin	83
Highly arched palate	96
Premature eruption of teeth	57

Based on 80 cases reviewed by J. Jaeken *et al.* See Ref. 25. There is an obvious ascertainment bias towards the severe end of the phenotypic spectrum.

been inconclusive. Hyperthyroidism was observed in three patients and a fourth patient had Kocher-Debre-Semelaigne syndrome, characterized by hypothyroidism and muscular hypertrophy.[2]

Most cases occur sporadically. Concordant monozygotic twins have been reported three times. Eight families with dominant transmission are known.[2] On the basis of well documented cases, an X-linked dominant mode of transmission cannot be ruled out. Male-to-male transmission, which rules out X-linked inheritance, has been mentioned three times in *undocumented* cases. Some sporadic instances may represent dominant mutations. However, some cases thought to be sporadic may, in fact, have an affected parent who remains undiagnosed. The diagnosis of Sotos syndrome in adults can be difficult.[26] First, advanced bone age is no longer a diagnostic aid. Second, final height attainment in not strikingly increased. Third, intellectual deficits tend to be mild. Fourth, the dysmorphic features may not be particularly striking. Finally, there seems to be a lack of awareness of Sotos syndrome by most internists.

Affected sibs of possibly consanguineous parents are well documented,[27] as well as the occurrence of Sotos syndrome in three sibs: monozygotic twin girls and their brother.[28,29] Thus, there appears to be evidence for an autosomal recessive form of Sotos syndrome, although parental gonadal mosaicism of a dominant gene cannot be ruled out.

Several other familial instances have been noted: two affected brothers whose maternal grandfather and maternal uncle had macrocephaly as an isolated finding,[30] Sotos syndrome in first cousins[31] and affected third cousins.[32]

NEVO SYNDROME

Some of the features of the Nevo syndrome[33] are similar to those found in Sotos syndrome, including intrauterine overgrowth, accelerated osseous maturation, dolichocephaly, large extremities, clumsiness, awkwardness, and retarded motor and speech development. However, features of the Nevo syndrome not found in Sotos syndrome are generalized edema at birth, severe muscular hypotonia, contractures of the feet, and wrist drop with clinodactyly. Large, low-set, malformed ears and cryptorchidism were observed in some patients. Autosomal recessive inheritance has been established. Although the Nevo syndrome has been confused with Sotos syndrome, the overall pattern of anomalies is at variance with the diagnosis of Sotos syndrome and it represents a distinct entity.[34]

RUVALCABA–MYHRE SYNDROME

Ruvalcaba and Myhre[35] reported two sporadic instances of a recurrent pattern syndrome with some similarities to Sotos syndrome. Both affected males had a birthweight of over 4,000 grams. Pertinent findings included large head, widely spaced eyes, triangular face, mild mental retardation, intestinal polyposis especially of the colon, and pigmentary spotting of the shaft of the penis. One had skin lesions resembling acanthosis nigricans. Both patients were adults. No malignant tumors had been detected at the time of the report. Although this syndrome has some features in common with Sotos syndrome and other features somewhat suggestive of Peutz-Jegers syndrome (except for location of the lesions), the condition is at variance with both of these diagnoses and represents, in my opinion, a newly recognized entity.

COMMENTS ON THE SOTOSOID PHENOTYPE

Although Sotos syndrome, Nevo syndrome, and Ruvalcaba-Myhre syndrome represent different entities, and Sotos syndrome itself may possibly be etiologically heterogeneous, certain features of the Sotosoid phenotype may share a similar pathogenesis. Furthermore, there remain a number of

TABLE 5
Comparison of Features of Two Patients with the Weaver Syndrome

	Patient 1	Patient 2
Growth		
Excessive growth of prenatal onset	+ + +	+ + +
Accelerated osseous maturation	+ + + +	+ + + +
Performance		
Hypertonia	+ +	+
Hoarse, low-pitched cry	+ +	+ +
Developmental delay	?	?
Excessive appetite	+ +	+ +
Craniofacial		
Wide bifrontal diameter	+ + +	+ + +
Flat occiput	+	+
Large ears	+ +	+ +
Ocular hypertelorism	+ +	+ +
Long philtrum	+ +	+
Relative micrognathia	+	+
Limbs		
Hands		
Prominent finger pads	+ +	+ +
Simian crease	−	+
· Camptodactyly	+ +	+
Broad thumbs	+ +	+
Thin, deep-set nails	+ +	+ +
Feet		
Clinodactyly, toes	+	+
Talipes equinovarus	+ +	−
Short fourth metatarsals	+	−
Limited early elbow and knee extension	+	+
Widened distal femurs and ulnas	+ +	+ +
Skin		
Excessive, loose skin	+ +	+ +
Inverted nipples	+	+
Thin hair	+	+
Other		
Umbilical hernia	+ +	+
Inguinal hernias	+ +	−

+ through + + + +, present in varying degrees of severity; −, absent; ?, uncertain.

TABLE 6
Pertinent Findings in the Elejalde Syndrome*

	Patient 1	Patient 2
Growth		
Gigantism	+	+
Abdominal/Genitourinary		
Omphalocele	+	+
Accessory spleen	−	+
Megaureter	+	+
Megabladder	+	+
Megavagina	−	+
Redundant connective tissue	+	+
Proliferation of perivascular nerve fibers	+	+
Craniofacial		
Craniosynostosis	+	+
Ocular hypertelorism	+	+
Epicanthic folds	+	+
Downslanting palpebral fissures	+	+
Hypoplastic nose	+	+
Rudimentary auricles	+	+
Limb		
Short limbs	+	+
Polydactyly	−	+
Other		
Redundant neck skin	+	+
Hypoplastic lungs	+	+

*Based on data from B.R. Elejalde *et al.* See Ref. 38.

patients who are said to be Sotos-like who have either (a) some highly unusual characteristics, or (b) not enough features of Sotos syndrome to make the diagnosis with certainty. Such patients may also share a similar pathogenesis.

Growth stimulating factors regulated or secreted by the hypothalamus should be considered. Sotos[24] has observed that patients with craniopharyngiomas exhibit catch-up growth following surgery, having rapid growth velocity and developing pointed chins and large hands and feet somewhat resembling cerebral gigantism. Such patients have growth hormone deficiency and no clear cause for this type of overgrowth has yet been identified; it may be hypothalamic. If the Sotosoid phenotype is viewed as a defect of the hypothalamus, it could be considered nonspecific, and might occur as an isolated defect or with broader patterns of anomalies making up various syndromes.

WEAVER SYNDROME

In 1974 Weaver *et al.*[36] reported a new overgrowth syndrome in two patients (Table 5). Excessive overgrowth was of prenatal onset and continued postnatally at two to three times the normal rate with accelerated osseous maturation. One other case was published in 1979.[37] I am aware of five other instances of this striking condition. One presumed case may be associated with neuroblastoma, but this needs confirmation.

To date, all cases are sporadic, although in one of the unpublished cases the mother apparently has some of the characteristic features. The etiology is unknown and the syndrome needs to be further delineated.

ELEJALDE SYNDROME

In 1977 Elejalde *et al.*[38] described a spectacular overgrowth syndrome Birthweights in two patients were 7,500 and 4,300 grams. Features are summarized in Table 6. The condition has autosomal recessive inheritance.

The three known cases were incompatible with life. Excess connective tissue was found throughout the body except for the central nervous system. It was most prominent subcutaneously in the media of vessels, in the walls of the viscera, and interstitially in organs such as the pancreas and kidneys. Fibroblasts from Elejalde syndrome patients completed the whole cell cycle in 63% of the normal cell cycle time. Perivascular proliferation of nerve fibers was found in many viscera, especially the spleen, thymus, colon, heart and adrenal glands.

ACKNOWLEDGMENTS

This paper is dedicated to the memory of David W. Smith. His teachings about dysmorphic growth and development live on in the many clinicians who were influenced by him.

Parts of this chapter appeared in the book Associated Congenital Anomolies, Chapter 8, "Overgrowth Syndromes", edited by M. El Shafie and C.H. Klippel, published by Williams and Wilkins, Baltimore, 1981.

REFERENCES

1. Smith DW. Growth and its disorders. Philadelphia: W.B. Saunders Co., 1977.

2. Cohen MM Jr. Overgrowth syndromes. *In*: El Shafie M, Klippel CH, eds. Associated congenital anomalies. Baltimore: Williams and Wilkins Co., 1981: 71-104.

3. Irving I. The E.M.G. syndrome (exomphalos, macroglossia, gigantism) *In*: Rickham RP, Hacker WC, Prevolt J, eds. Progress in pediatrics. Vol. 1. Munich: Urban and Schwarzenberg, 1970: 1-61.

4. Fraumeni JF Jr. Stature and malignant tumors of bone in childhood and adolescence. *Cancer* 1967; 20: 967-73.

5. Tjalma RA. Canine bone sarcoma: Estimation of relative risk as a function of body size. *J Natl Cancer Inst* 1966; 36: 1137-50.

6. Horger EO, Facog M, Miller C, Conner ED. Relation of large birthweight to maternal diabetes mellitus. *Obstet Gynecol* 1975; 45: 150-54.

7. Fee SA, Weil WB Jr. Body composition of infants of diabetic mothers by direct analysis. *Ann Natl Acad Sci* 1963; 110: 869-97.

8. Naeye RL. Infants of diabetic mothers: A quantitative, morphologic study. *Pediatrics* 1965; 35: 980-88.

9. Vohr BR, Lipsitt LP, Oh W. Somatic growth of children of diabetic mothers with reference to birth size. *J Pediatr* 1980; 97: 196-99.

10. Cornblath M, Schwartz R. Disorders of carbohydrate metabolism in infancy. Philadelphia: WB Saunders Co., 1976.

11. Hansson G, Redin B. Familial neonatal hypoglycemia. *Acta Paediatr (Stockh.)* 1963; 52: 145-52.

12. Beckwith JB. Macroglossia, omphalocele, adrenal cytomegaly, gigantism, and hyperplastic visceromegaly. *Birth Defects*. 1969; 5(2): 188-96.

13. Wiedemann HR. Complexe malformatif familial avec hernia ombilicale et macroglossie – un syndrome nouveau? *J Genet Hum* 1964; 13: 223-32.

14. Sotelo-Avila C, Gonzalez-Crussi F, Fowler JW. Complete and incomplete forms of the Beckwith-Wiedemann syndrome: Their oncogenic potential. *J Pediatr* 1980; 96: 47-50.

15. Kosseff AL, Herrmann J, Gilbert EF, Viseskul C, Lubinsky M, Opitz JM. The Wiedemann-Beckwith syndrome: Clinical, genetic and pathogenetic studies of 12 cases. *Eur J Pediatr* 1976; 123: 139-66.

16. Herrmann J, Opitz JM. Delayed mutation as a cause of genetic disease in man: Achondroplasia and the Wiedemann-Beckwith syndrome. *In*: Nichols WW, Murphy DG, eds. Regulation of cell proliferation and differentiation. New York: Plenum Press, 1977.

17. Gardner LI. Pseudo-Beckwith-Wiedemann syndrome: Interaction with diabetes mellitus. *Lancet* 1973; 2: 911-12.

18. Berry AC, Belton EM, Chantler C. Monozygotic twins discordant for Weidemann-Beckwith syndrome and the implications for genetic counseling. *J Med Genet* 1980; 17: 136-38.

19. Cohen MM Jr. The child with multiple birth defects. New York: Raven Press, 1982.

20. Gorlin RJ, Meskin LH. Congenital hemihypertrophy. *J Pediatr* 1962; 61: 870-79.

21. Parker DA, Skalko RG. Congenital asymmetry: Report of ten cases with associated developmental abnormalities. *Pediatrics* 1969; 44: 584-89.

22. Gorlin RJ, Pindborg JJ, Cohen MM Jr. Syndromes of the head and neck. 2nd ed. New York: McGraw-Hill Co., 1976.

23. Noe O, Berman HH. The etiology of congenital hemihypertrophy and one case report. *Arch Pediatr* 1962; 79: 278-88.

24. Sotos JF, Cutler EA. Cerebral gigantism *Am J Dis Child* 1977; 131: 625-27.

25. Jaeken J, van der Schueren-Lodeweycky M, Exckels R. Cerebral gigantism syndrome. *Z Kinderheilkd* 1972; 112: 332-46.

26. Zonana J, Sotos JF, Romshe CA, Fisher DA, Elders MJ, Rimoin DL. Dominant inheritance of cerebral gigantism. *J Pediatr* 1977; 91: 251-56.

27. Boman H, Nilsson D. Sotos syndrome in two brothers. *Clin Genet* 1980; 18: 421-27.

28. Townes PL. Cerebral gigantism. *J Med Genet* 1976; 13: 80.

29. Townes PL, Scheiner AP. Cerebral gigantism (Sotos syndrome): Evidence for recessive inheritance. *Pediatric Res* 1973; 7: 349.

30. Bejar RL, Smith GF, Park S, Spellacy WH, Wolfson SL, Nyhan WL. Cerebral gigantism: Concentrations of amino acids in plasma and muscle. *J Pediatr* 1970; 76: 105-11.

31. Hooft C, Schotte H, Van Hooren F. Gigantisme cerebral familial. *Acta Paediatr (Belg.)* 1968; 22: 173-86.

32. Krauel X, Berger R, Amiel-Tison C. Gigantisme cérébral: Deux cas familiaux. *J Genet Hum* 1977; 25: 205-14.

33. Nevo S, Zeltzer M, Benderly A, Levy J. Evidence for autosomal recessive inheritance in cerebral gigantism. *J Med Genet* 1974; 11: 158-65.

34. Cohen MM Jr. Diagnostic problems in cerebral gigantism. *J Med Genet* 1976; 13: 80.

35. Ruvalcaba RHA, Myhre S, Smith DW. Sotos syndrome with intestinal polyposis and pigmentary changes of the genitalia. *Clin Genet* 1980; 18: 413-16.

36. Weaver DD, Graham CF, Thomas IT, Smith DW. A new overgrowth syndrome with accelerated skeletal maturation, unusual facies, and camptodactyly. *J Pediatr* 1974; 84: 547-52.

37. Shimura T, Utsumi Y, Fujikawa S, Nakamura H, Bata K. Marshall-Smith syndrome with large bifrontal diameter, broad distal femora, camptodactyly, and without broad middle phalanges. *J Pediatr* 1979; 94: 93-5.

38. Elejalde BR, Giraldo C, Jimenez R, Gilbert EF. Acrocephalopolydactylous dysplasia. *Birth Defects* 1977; 13(3B): 53-67.

COUNSELING IN CASES OF IDIOPATHIC SYNDROMES

John M. Opitz

INTRODUCTION

This presentation will be short on genetic counseling but long on the biology of syndromes. In view of the extraordinary confusion that exists on the subject "syndrome" in clinical genetics, it will be necessary to comment on the recent efforts by an International Working Group (IWG) on Concepts and Terminology of Errors of Morphogenesis whose report and recommendations are at last in press.[1] First let us consider the counseling aspects of this complex topic.

COUNSELING CONSIDERATIONS

The bane, and to some extent the most interesting aspect, of my professional existence is the hundreds of mail inquiries on undiagnosed cases I receive annually and the many cases of "unknown syndromes" that are presented to me wherever I go. The following three questions are always the same: What is it? What is the prognosis? What is the cause and recurrence risk? In what follows I will give a few hints on how to approach the first question; to a large extent the answer to the second question depends on how skillfully the first is answered. I would estimate that in only 20% of such cases am I able to answer the third question on cause and recurrence.

In the case of the "funny looking kid" (FLK) it is the doctor, not the mother, who is confused. Long evolutionary experience of recognizing their own child at a great distance in a crowd of other children (or of learning to tell their identical twin babies apart!) has made mothers wise about the analysis of normal developmental variation and family resemblance, and they are usually able to tell the consulting clinical geneticist whether their child is "normal" or not, *i.e.*, whether the child needs a chromosome analysis or not. Often it is the mother who first suspects Down syndrome (DS). Not too long ago in Wisconsin most birth certificates of known DS children still read "normal newborn male/female". It is a distressing phenomenon that upon entry into medical school most sons and daughters of such normally astute mothers precipitously loose their ability at phenotype analysis. The matter is not helped by the fact that normality is anathema and generally not taught in medical school except in terms of standard deviations and other such statistical abstractions of biology.

The most common counseling problem presented to me is the child with the idiopathic sporadic true multiple congenital anomalies-mental retardation (MCA/MR) syndrome, *i.e.*, no diagnosis can be made, parents are not consanguineous, family history is unremarkable and the child's chromosomes are normal. In such cases the empiric recurrence risk is less than 1%*. Occasionally I do a chromosome study on such parents on the off chance that their affected child may have inherited from one of them a virtually undetectable translocation chromosome. I have never detected a translocation heterozygote in this manner yet.

Many more empiric recurrence risk data are needed before we can counsel with greater confidence in cases of idiopathic sporadic MCA syndromes. Long term follow-up is required before such data will yield an answer to the question as to whether such an empiric recurrence risk figure, say of 4.2%, applies uniformly to the entire cohort or represents the weighted mean of at least two populations: a large one with an essentially "zero" recurrence, and a minority with "high" recurrence of say 25% on the average. Given our luck and the biologic nature of our material the truth may ultimately turn out to be even more complex than that. The classification criteria and descriptive nomenclature review which follows further illustrates these problems.

FIELDS AND MALFORMATIONS

There is of course more to syndromes than just malformations; however, no concept is as vital to an understanding of syndromes in clinical genetics than that of primary malformation and the (embryological) developmental field.

The IWG defined a malformation as a "morphological defect of an organ, part of an organ or a larger region of the body, resulting from an intrinsically abnormal developmental process". The emphasis is on organogenesis, not histogenesis. Most malformed organs are histologically normal, but it is quite possible that an embryonic defect of histogenesis might on occasion lead to a defect of morphogenesis. The definition refers to "*intrinsically* abnormal processes", *i.e.*, mostly mutational and to some extent multifactorial defects which destine the primordium involved to be abnormal from conception. This defines a *primary* malformation. A *secondary* malformation, now being referred to as a disruption, was defined by the IWG as "morphologic defect of

---

*Gilbert JD, A review of 151 cases of idiopathic, sporadic mental retardation with multiple congenital anomalies: empiric recurrence risk. Research report submitted in partial fulfillment of requirements for the degree of Master of Science (Medical Genetics) University of Wisconsin. 1974.

of an organ, part of an organ, or larger region of the body resulting from the breakdown of, or an interference with an originally normal developmental process". Here the primordium was initially destined to be normal, but was disrupted during the course of its development through an *extrinsic* agent or event such as a teratogenic drug, physical agent such as hypothermia or radiation, metabolic disturbance such as maternal diabetes or phenylketonuria (PKU), infectious agent such as rubella, cytomegalo or herpes virus or toxoplasmosis, endocrine disturbance such as maternal arrhenoblastoma masculinizing a daughter, or maternal thyroid antibody interfering with fetal thyroid development, *etc*.

As you know, teratogenic phenocopies may look identical to malformations produced by mutation: anatomically a triphalangeal thumb produced by thalidomide probably would be identical to the triphalangeal thumb due to an autosomal dominant radius aplasia mutation. In other words, primary and secondary malformations both affect developmental fields. In practice a useful distinction is made between the two by saying a malformation is a defect *of* a field, a disruption occurs *in* a field.

What then is the developmental field? A developmental field is defined as a part of the embryo in which the processes of development of the complex structure appropriate to that part are controlled and coordinated in a spatially ordered, temporally synchronized, and epimorphically hierarchial manner.

Normal anatomical structure of course is mute about events which shaped it. Nevertheless it must not be supposed that the anatomical components, say of a thumb, attained their definitive location through random events affecting each part separately. Spatial ordering refers to the processes whereby parts of a complex structure assume their proper location and orientation to each other through a combination of genetically determined active and passive morphogenetic movement, growth, and cell-cell mediated, mutually inductive relationships. Thus failure of the normal, mutually inductive relationships between midface and forebrain may lead to synophthalmic cyclopia with the absence of medial portions of orbita, ethmoid sinuses, crista galli and cribriform plates, proboscis on forehead together with holoprosencephaly and premaxillary agenesis. An occasional "experiment" of nature gives a vivid insight into such a mutually inductive relationship as in the mentally retarded boy reported by Jones, *et al*,[2] whose left eye was located on the side of the head where it had induced a complete orbit with extraocular muscles and lachrymal glands, lids, lashes, an area free of scalp hair around it and an eyebrow.

Temporal synchronization in a field refers to the timing and rate of simultaneously occurring morphogenetic events so that failure of one part to develop will retard another, as may occur in cases of failure to break down interdigital webs with resulting shortness and incomplete development of digits. Thus, control of timing also affects the numerous processes of degeneration which are an important part of normal development and responsible for the disappearance of the tail and interdigital webs, perforation of anus, *etc*.

The epimorphically hierarchial aspects of the field refer to its sequential developmental progression from a less to a more highly complex and mature stage along the lines of the table of progressive differentiation in every embryology textbook. It gradually limits and reduces the size of field in which a dysmorphogenetic cause can act. Thus, early in limb development a mutation may produce a phocomelia (a disturbance of epimorphic progression in which incomplete rhizomelic development proceeds to imperfect acromelic development without an intervening stage of mesomelic development). A bit later, the same cause may produce in a different individual a radius aplasia, and even later a triphalangeal thumb or only a cleft of the distal phalanx of the thumb.

During initial stages of development the pluripotential ovum is referred to as the primary field; progressive differentiation causes individual regions to be irreversibly determined and to become epimorphic fields. Initially all field control and integration is probably attained locally through mutually inductive relationships and messages. Later, long distance control may influence development as through the effect of the circulation on cardiac morphogenesis of testosterone, dihydrotestosterone and the anti-Müllerian hormone on male genital development, and the postulated effect of neural crest mediated sensory innervation on limb development.

Most field components are closely contiguous and are referred to as *monotopic* fields: holoprosencephaly being one example, preaxial polydactyly another. In other fields the components are more distantly located and are called *polytopic*. An example is the acrorenal field inferred from correlated occurrence of renal and acral limb malformations and experimental work which shows that mesonephros is required for limb cartilage to differentiate *in vitro*.

Evidence for the existence of fields comes from clinical genetics, teratology, developmental genetics, comparative anatomy and embryology, experimental embryology, and stochastic syntropology and has been reviewed recently.[3-6] The most powerful argument is from clinical genetics, the subject most accessible to most of us. It asserts that a field has been defined when an anatomically identical developmental defect, regardless of how extensive or complex, is found to be caused in different individuals by two or more different causes. This is because identical structure means identical development, regardless of difference in initial causal mechanism. Human, and for that matter all mammalian organs, are evidently capable of responding to a huge number of diverse morphogenetic causes with the production of only a very limited repertoire of malformations precisely because coordinated field development limits independent response from component structures. Therefore, there is no such thing as a causally specific malformation.

The phylogenetic origin of field development must be seen in the gradual reduction of metamerism during vertebrate evolution. Of course a considerable amount of metamerism persists in adult structure of vertebrae, ribs and intercostal structures, and appears transiently in embryonic life as in branchial arches, pronephros, and metanephros, *etc.* Field development means coordinating previously less complex multiple metameric structures into single more complex structures, such as the six branchial arches into a single jaw-neck system integrated to a large extent by neural crest cells, and its six pairs of aortic arches into a single aortic arch system. Ontogenetic beginnings or potential may be revealed prenatally in vestiges, *i.e.* persistence or incomplete regression of an embryologic structure such as a Meckel diverticulum as a proximal remnant of the omphalomesenteric duct. Atavisms ("appearance of structures presumed to have been present in a remote ancestor") are a powerful argument for developmental homology and have been seen particularly convincingly in many muscle abnormalities in human aneuploidies.

MILD MALFORMATIONS VERSUS MINOR ANOMALIES

Until recently this distinction was not made clearly, if at all, yet it is of supreme importance especially when trying to decide whether an aneuploidy syndrome is present or not.

In the view of Waddington[7,8] the major developmental events which lead to discontinuous differences in morphological structure in the mature organism are threshold decisions. Most primary malformations represent two major classes of threshold defects:

1. A threshold towards completion of the structure was not crossed. These are the *defects of incomplete differentiation* and are the commonest malformations (*e.g.* hypospadias, cleft palate, spina bifida). The ageneses and aplasias and transverse terminal arrests of limb development are probably a variant of this class but may, in part, represent an earlier effect. Or,
2. The underlying defect forced a developmental process into an unnatural path. These are the less common malformations, namely the *defects of abnormal differentiation* (*e.g.* extra thumb). Thus malformations, including their least severe forms, are all-or-none traits, *i.e.* they are neither graded, nor metric, and at their least severe end they do not shade into normality. Meaning, mild malformations do not occur as normal variants in a population.

However, minor anomalies cannot be distinguished from, and developmentally are identical to, normal developmental variants which are the numerous graded, anthropometric variations of final structure and their rate of attainment. Normal anthropometric variants are the finer details put on our features during the last stages of development and are the traits which constitute our morphologic uniqueness, but which are also the heritage of ethnic group and of family inheritance. To use a homely metaphase, normal development variants are the results of end-stage "fine tuning" of development and do not involve major threshold decisions of morphogenesis. Almost by definition these are polygenic traits; no mendelian minor anomalies are known. By contrast, oligogenes play major roles in morphogenetic threshold events, and, however infrequently, most malformations have been found as mendelian traits (*e.g.* X-linked cleft palate, X-linked imperforate anus, X-linked or autosomal dominant spina bifida). All normal developmental variants can occur as minor anomalies and vice versa, but while developmentally identical they have different causal implications. Every textbook on physical anthropology contains numerous examples of normal developmental variants, the distribution of their measurements in the population by age, sex and race, and an analysis of their respective genetic and environmental contributions to the cause of their variability. Whole textbooks have been written on dermatoglyphics alone, a particularly useful set of traits because they can be quantitated so exactly from good prints and they remain uninfluenced by age or internal environment after 12 weeks of development. All dermatoglyphic traits in DS for example (high number of ulnar loops, increased incidence of radial loops on fingers four and five, and distal loops in the third interdigital area, distal axial triradii and increased *atd* angles, and hallucal open fields) are commonly seen individually in the general population as normal variants. In DS they are minor anomalies and collectively are so characteristic as to be diagnostic. Aneuploidy disturbs preeminently the end stage fine tuning of development as is evident from the well-known fact that most anomalies in such conditions are minor anomalies and such individuals lack family resemblance. Asking mothers and grandmothers whether the child with a syndrome has family resemblance can be an exceedingly sensitive test of presence or absence of aneuploidy.

In forensic medicine an analysis of such developmental variants may aid in identifying race. In clinical genetics this analysis serves to differentiate an FLK with multiple normal but perhaps somewhat unusual developmental variants (normal for its family) from a child with a bona fide syndrome of multiple minor anomalies (not normal for the family) and perhaps needing a buccal smear or chromosome analysis. Thus a study of minor anomalies teaches three important lessons:

1. Learn the normal, normality being infinitely more variable than abnormality.
2. Learn how to measure and make as many quantitative statements as possible about the child's anomalies.
3. Examine relatives; normal variants are familial, syndromic minor anomalies usually are not. Pleiotropic mendelian mutations usually do not produce syndromes of multiple minor anomalies.

DEFORMITIES

The IWG defined a deformation as an "abnormal form, shape or position of a part of the body caused by . . . mechanical forces". Deformities may be of extrinsic or intrinsic origin, e.g. oligohydramnios due to chronic amniotic fluid leak may lead to the Potter syndrome (extrinsic in origin), severe myotonic dystrophy may lead to an arthrogryposis-like set of deformities (intrinsic in origin). In both of these cases the onset is congenital. In a prenatally brain damaged child with subsequent cerebral palsy the later plagiocephaly, kyphoscoliosis, dislocated hips and flexion deformities of limbs are postnatally acquired deformities. Skeletal dysplasias may be associated with deformities of pre- and/or postnatal onset, e.g. osteogenesis imperfecta. Children with hemivertebrae generally have or will develop scoliosis and some metabolic disorders may result in deformities, e.g. bowing of legs in hypophosphatemic rickets. Other deformities may represent the effect of dysplasia (scoliosis in neurofibromatosis).

The biology and pathogenesis of deformities are usually quite easy to discern; they are a common component of mendelian and aneuploidy syndromes.

DYSPLASIAS

The IWG defined dysplasia as an ". . . abnormal organization of cells into tissue(s) and its morphologic result(s)". Thus a dysplasia is due to dyshistogenesis, i.e. represents a defect of tissue differentiation (in contradistinction to a malformation which represents a defect of dysmorphogenesis).

In general the biology of dysplasias is well known to clinical geneticists; a few summarizing statements follow:

1. Dysplasias may represent metabolic defects (as in fibrositis ossificans progressiva, the Zellweger syndrome, the acid mucopolysaccharidoses, the mucolipidoses, etc.), most are non-metabolic defects.
2. They may involve derivative(s) of one (colonic polyposis) or two or more germ layers (tuberous sclerosis).
3. They may be generalized (as in a skeletal dysplasia) or localized (presacral teratoma).
4. They may be single lesions (single acoustic neuroma) or consist of multiple lesions (von Recklinghausen neurofibromatosis).
5. They are usually benign lesions (e.g. encephalotrigeminal angiomatosis), but may be premalignant (as in colonic polyposis).
6. They are usually permanent lesions (giant pigmented hairy nevus), but some are evanescent (cavernous hemangioma, neuroblastoma type IV).
7. They may be congenital (cutis marmorata telangiectatica congenita, presacral teratoma) or postnatal (testicular teratocarcinoma) in onset.
8. They may be disruptive or nondisruptive on development.

9. They may be dystopic (brain tissue in lung of an anencephalic) or non-dystopic (retinoblastoma).

Many dysplasias can be viewed as developmental fine-tuning defects (*vide supra*) of histogenesis. Small, minor dysplasias are exceedingly common in the normal population and include torus palatinus and mandibularis, iris freckling, capillary hemangioma over glabella, metopic suture and nape of neck, ephelides, an occasional café-au-lait spot, moles, nevi, fibromata, papillomata, an occasional lipoma, some spider angiomata (pregnant women), Campbell-de Morgan angiomata, *etc.* Thus dysplasias are causally non-specific. The neural crest dysplasia, multiple neurofibromata, can occur as an autosomal dominant trait (von Recklinghausen's disease), and as a component manifestation of the Noonan syndrome.

Most dysplasias occur sporadically and are presumed to be multifactorially caused. If they are mendelian traits, they mostly represent autosomal dominant mutations.

Dysplasias are components of every aneuploidy syndrome and are the probable reason for the increased incidence of cancers in them (*vide infra*).

Dysplasias may be induced environmentally, either prenatally (*e.g.* maternal DES treatment with dystopic vaginal adenosis in their daughters) or postnatally through radiation, viruses, carcinogens, *etc.*

OTHER TERMS

The IWG also agreed on the following definitions:

Hypoplasia, hyperplasia: Under- and overdevelopment of an organism, organ or tissue due to a decrease or increase of the number of cells respectively.

Hypotrophy, hypertrophy: Decrease and increase in the size of cells, tissue, or organ respectively. Three well-known classes of hypotrophy include neurohypotrophy, ischemic hypotrophy, endocrine hypotrophy.

Agenesis: Absence of a part of the body due to an absent primordium.

Aplasia: Absence . . . due to failure of primordium to develop.

Atrophy: A normally developed mass of tissue(s) or organ(s) decreases due to a reduction in cell size and/or number.

TERMS AND CONCEPTS FOR PATTERNS OF MULTIPLE DEFECTS

Congenital anomalies are frequently associated with other anomalies in the same person as recognizable or puzzling patterns and the most discerning analysis may be necessary to understand the causal and pathogenetic relationship between them.

The word "syndrome" is frequently applied to such patterns regardless of the level of understanding. In a general way, such understanding ultimately comes about on a pathogenetic and/or causal level, with the one not necessarily implying the other, *i.e.* in the Ellis-van Creveld syndrome cause is quite well understood, but pathogenesis is not. In the Potter syndrome pathogenesis is reasonably well known, but the cause for the urethral obstruction or bilateral renal agenesis is frequently not evident.

To aid understanding on a pathogenetic level the IWG introduced the term *sequence* which was defined as ". . . multiple defects derived from a single known or presumed prior structural or mechnical factor". This exceedingly useful term can be applied to a number of situations.

1. The consequences of a primary malformation, *i.e.* the dislocated hip, rectal and urinary incontinence, clubfeet, and neurohypotrophy of lower limbs and their dermatoglyphics as a consequence of spina bifida (the whole clinical picture simply being called spina bifida sequence), similarly the Ebstein anomaly sequence, the Pierre Robin anomaly sequence, the bilateral renal agenesis sequence, *etc.*
2. The consequence of a teratogenic effect in a field is in general terms called a disruption sequence, *e.g.* the thalidomide disruption sequence (or thalidomide sequence for short), irradiation disruption sequence, amniotic deformities-adhesions-mutilations (ADAM) sequence, *etc.* Note one important difference in the way the word is used in primary and in secondary malformations: in the former it does not include the pathogenesis of the malformation, in the latter it does. The pathogenesis of primary malformations is subsumed under the concept of developmental field defects.
3. The consequences of action of mechanical factor (oliogohydramnios, uterine pressure on weakened limbs) is referred to as a deformation sequence, *e.g.* Potter sequence, arthrogryposis sequence, *etc.*
4. The secondary effects of a dysplasia are dysplasia sequences, *e.g.* the Kassabach-Merritt sequence in a cavernous hemangioma; the renal, pulmonary, CNS and cardiac sequences in tuberous sclerosis, *etc.*

The IWG defined the word *syndrome* as a ". . . pattern of multiple anomalies thought to be pathogenetically related and not representing a sequence". This understanding of the concept straddles pathogenetic and causal aspects, because the language of the definition does not preclude its use for a causally defined entity such as the Ellis- van Creveld, 18-trisomy, thalidomide or fetal alcohol syndrome. It suggests the possibility that the term may be applied unintentionally to the complex manifestations of a single (perhaps polytopic) developmental field defect as yet not recognized as such. In this manner some workers refer to the Curran syndrome[9,10] rather than to the acrorenal field defect (or complex), the Poland syndrome rather than the Poland anomaly (or complex, or field defect).

What about concurrence of anomalies not known to be a field defect, sequence or syndrome? Here the IWG addressed itself to the one meaning of the term *association* as recently used in clinical genetics.* Since there are no obligatory anomalies in malformation syndromes, *i.e.* anomalies present in 100% of cases (or required for diagnosis), then the observation of anomaly A being associated in 100% of cases with anomaly B invites interpretation. After studying the matter, B may be found to be a symptom of A (cyanosis or clubbing in tetralogy of Fallot, MR in the DS, *etc.*), or B may be a component of a field defect (cylopia in holoprosencephaly). If B is seen in 60% of the cases of A then B may be a facultative effect or manifestation of A, or A and B may be pleiotropic manifestations of one cause, or may represent a looser type of noncausal association (*e.g.* liability to midline defects as documented by Czeizel in his schisis associations).[12]

Mindful of these considerations, the IWG defined an *association* as ". . . non-random occurrence in one or more individuals of multiple anomalies not known to be a field defect, sequence or syndrome". Current examples are the VATER association, Müllerian agenesis, aortic arch and vertebral defects associations, *etc.* If approached from a statistical or epidemiological point of view, malformation associations are nothing more than stochastic entities (*i.e.* arbitrary truncations of an infinite chain of associations) and not biological entities. The hazards and pitfalls of proceeding to define syndromes in this manner are numerous, especially if begun in collections of university hospital material which concentrates with highest likelihood on the rarest of concurrences – or of specialty clinics such as pediatrics, cardiology, or MR whose patient populations may suggest that all patients with B have A (*e.g.* congenital heart defects or MR). Nevertheless, the establishment of associations has one important benefit in clinical practice, because knowing the association of A with B in x%, with C in y%, and with D in z% of cases, *etc.*, alerts the clinician to look for B,C, and D in all cases of A. Thus an infant with feeding problems and pulmonary aspiration who is found to have thoracic vertebral anomalies should be checked for a TE fistula, and is also more likely to have CHD, renal anomalies, *etc.* What is the ultimate fate of associations? Some may turn out to be one or more bona fide causally defined syndromes, some complex polytopic fields, and some only chance concurrences.

For most clinical geneticists the most effective and valid way to define syndrome is the biological/genetical way whereby phenotype analysis in the first patient with a given combination of anomalies may suggest pathogenetic and causal hypotheses, and therefore encourage a search for additional cases, for etiologic information in the family history, chromosome examination, or closer scrutiny of relatives. The causally most securely defined syndromes are

---

*The term also has an anterior, different meaning as used by immunogeneticists studying blood group or HLA-disease associations.

those with a chromosome anomaly (where syndrome definition – but by no means delineation – may be accomplished already in the first patient studied), or those in which chromosomes are normal but in which sib and/or family occurrence identifies a segregating mendelian mutation. This process of syndrome definition was best articulated by Möbius, who wrote in 1892:[13] "In elucidating . . . diseases, the student of human pathology proceeds from the observation of one patient. The patient presents the initial picture of the condition. The same or a similar picture is also observed in other patients. Astute observation recognizes what is identical in different patients; new pathological concepts are obtained by differentiation between coincidental and essential and non-essential findings. In this manner one initially obtains a symptomatic entity. . . Only after some time does observation on natural history and anatomical findings succeed in separating symptomatic similarities and in uniting differences. The process ends only when, in addition to the symptoms, the natural history, the anatomical findings, *etc.* the cause . . . has been recognized, *i.e.* when the disease has become an etiologic entity."

At the level of the first patient one speaks of a "physical examination syndrome", at the level of the symptomatic entity one speaks of a "formal genesis syndrome", and at the level of the etiologic entity as a "causal genesis syndrome".[14] In cases of the Hallermann-Streiff and the Rubinstein-Taybi syndrome the causal genesis level has not yet been reached; in the Prader-Willi syndrome it seems to have been attained with the observation of the proximal 15q chromosome abnormalities.[15] In the case of many single major anomalies and the Williams hypercalcemia metabolic dysplasia syndrome, a somewhat indeterminate level of understanding has been reached by postulating a normally distributed polygenic predisposition with a variable threshold as cause and generally low empiric recurrence risk.

In the case of causally defined syndromes, the multiple manifestations of the patients are referred to as pleiotropy. Two general forms of pleiotropy have been defined. In relational pleiotropy all manifestations can be shown to relate to one underlying basic defect probably expressed on an intercellular basis, *e.g.* the dislocation of lens, skeletal, and cardiovascular anomalies in the Marfan syndrome relating to a basic connective tissue defect. All manifestations in the Elejalde syndrome are thought to relate to a defect in regulation of cell division,[16] in the Zellweger syndrome to a defect of energy metabolism probably due to a defect of cytochrome b.[17] In mosaic pleiotropy it is difficult if not impossible to relate all manifestations to a single common connective tissue or metabolic defect. In most cases it is thought that the mutant gene exerts its effect on an intracellular basis at different times in the different abnormal derivatives. Thus in the Ellis-van Creveld syndrome the congenital heart defect, polydactyly, skeletal dysplasia and multiple buccal frenular abnormalities are difficult to relate to a single common factor disturbing morphogenesis in all of these organs.

SUMMARY OF BIOLOGICAL CONSIDERATIONS

Patterns of multiple anomalies can be grouped as follows:

I. Phenotypic Classification
   A. Sequences
      1. Developmental field defects (malformations) and malformation sequences
         a. monotopic: spina bifida sequence
         b. polytopic: acrorenal field defect
      2. Disruption sequences (*e.g.* ADAM defect, fetal alcohol sequence)
      3. Dysplasia sequences
         a. metabolic (*e.g.* Zellweger syndrome, fibrodysplasia ossificans progressiva)
         b. nonmetabolic (*e.g.* tuberous sclerosis sequence of MR, fits, pneumothorax, renal failure)
      4. Deformity sequences
         a. intrinsic (*e.g.* multiple congenital contractures due to severe myotonic dystrophy-mother affected)
         b. extrinsic (amniotic fluid leak→Potter deformity sequence)
   B. Malformation Association
      1. VATER association
      2. Schisis association
      3. CHARGE association
      4. Müllerian agenesis association with aortic arch and vertebral anomalies, *etc.*
   C. Syndromes*
   D. Consisting of variable combinations of:
      1. Malformations
      2. Dysplasias
      3. Deformities
      4. Minor anomalies
      5. Growth impairment
      6. Maturation defects

II. Causal Classification
   A. Mendelian mutations with
      1. mosaic or
      2. relational pleiotropy

---

*Except for some large molecular weight inborn errors of metabolism (*e.g.* Hunter, Sanfilippo syndrome), the inborn errors of metabolism, their manifestations, and sequelae are referred to as diseases, not syndromes.

   B. Chromosome abnormalities
   C. Multifactorial syndromes? (*e.g.* Williams' syndrome)
   D. Environmental causation (disruption sequences)
   E. Maternal fetal interaction
      1. maternal PKU
      2. maternal diabetes
      3. Williams' syndrome?
   F. Cause unknown (*e.g.* Associations, Rubinstein-Taybi syndrome)
   G. Coincidental syndrome.

Phenotype analysis is a prerequisite skill before causal statements should be attempted.

SUMMARY

Counseling in cases of idiopathic syndromes involves mostly correct phenotype analysis to determine the exact biological nature of the syndrome present. The IWG recently arrived at a concensus and a set of recommendations concerning concepts and terms of errors of morphogenesis (*J Pediatr*, in press). These address the embryonic developmental field and its intrinsic defects of morphogenesis leading to (primary) malformations and its extrinsically caused developmental disturbances resulting in disruptions. They also address the secondary and even later consequences on pre- and postnatal structure and function of the individual due to malformations, disruptions, deformities and dysplasia; hence, spina bifida sequence, thalidomide sequence, oligohydramnios sequence, *etc.*

Complex malformations, sequences and associations must be distinguished from syndromes (pattern of multiple, pathogenetically related anomalies not representing a sequence). In causally defined syndromes, *i.e.* Down syndrome, Ellis-van Creveld syndrome, the pathogenetic relationship between these multiple anomalies is subsumed under the concept of pleiotropy.

In analyzing syndromes it is important to keep in mind the distinction between the least severe forms of malformations and minor anomalies. Individually, the latter are phenotypically indistinguishable from the normal anthropometric developmental variants which form the morphologic basis of individuality, family resemblance and racial affinity. Most of the developmental disturbances in aneuploidy syndromes are minor anomalies, each syndrome being characterized by its own unique combination of many minor anomalies which enable diagnosis and abolish family resemblance.

Phenotype analysis is a necessary prerequisite of genotype analysis and counseling in syndromes. In idiopathic sporadic MCA/MR syndromes (without chromosome abnormality or parental consanguinity) the empiric recurrence risk is less than 2%. However, in view of newer, 1000 band prometaphase resolution methods, such studies need to be repeated because the counseling consequences for normal relatives carrying a minute balanced chromosome rearrangement are quite different from those for heterozygotes of an autosomal recessive gene causing the given MCA/MR syndrome.

REFERENCES

1. Spranger JW, Opitz JM, Smith DW, *et al*. Errors of morphogenesis: Concepts and terms; Recommendations of an international working group. *J Pediatr* 1981. (in press)
2. Jones KL, Higginbottom MC, Smith DW. Determining role of the optic vesicle in orbital and periocular development and placement. *Pediatr Res* 1980; 14: 703-708.
3. Gilbert EF, Opitz JM. Developmental and other pathological changes in syndromes due to chromosome abnormalities. *In*: Rosenberg HS, ed. Perspectives in pediatric pathology. Chicago: Year Book Medical Publisher, 1981. (in press)
4. Opitz JM, Gilbert EF. Pathogenetic analysis of congenital anomalies in humans. *In*: Ioachim HL, ed. Pathobiology annual. New York: Raven Press, 1981.
5. Opitz JM. The developmental field concept in clinical genetics. David W Smith Festschrift presentation Lake Wilderness Center, WA. *J Pediatr*, 1981. (in press)
6. Opitz JM. Topics IV,V. Terminology and pathogenetic analysis of human congenital anomalies. *In*: Tópicos Recentes de Genética Clínica. Editora Pedagógica Universitaria e Editora da Universidade de São Paulo, 1982. (in press)
7. Waddington CH. Canalization of development and the inheritance of acquired characteristics. *Nature* 1942; 150: 563-65.
8. Waddington CH. The cybernetics of development. *In*: Waddington CH, ed. The strategy of the gene. London: Allen and Unwin 1957.
9. Salmon MA, Wakefield MA. The acrorenal syndrome of Curran. *Dev Med Child Neurol* 1977; 19: 521-24.
10. Dieker H, Opitz JM. Associated acral and renal malformations. *BDOAS* 1969; V/3: 68-77.
11. Rosenberg LE, Kidd KK. HLA and disease susceptibility: a primer. *New Eng J Med* 1977; 297: 1060-62.
12. Czeizel A. Schisis associations. *Am J Med Genet* 1981. (in press)
13. Möbius PJ. Ueber infantilen Kernschwund Muenchn Med Wschr. 1892; 39: 1ff.
14. Opitz JM, Herrmann J, Pettersen JC, Bersu ET, Colacino SC. Terminological, diagnostic, nosological, and anatomical-developmental aspects of developmental defects in man. *In*: Harris K, Hirschhorn K, eds. Advances in human genetics. Vol. 9 New York: Plenum, 1979; 71-164.

15. Ledbetter DH, Riccardi VM, Airhart SD, Strobel RJ, Keenan BJ, Crawford JD. Deletions of chromosome 15 as a cause of the Prader-Willi syndrome. *New Eng J Med* 1981; 304: 325-29.

16. Elejalde BR, Giraldo C, Jimenez R. Gilbert EF. Studies of malformation syndromes in man XLIV. Acrocephalopolydactylous dysplasia: A previously undescribed autosomal recessive malformation/dysplasia syndrome. *BDOAS* 1977; XII/3B: 53-68.

17. Versmold HT, Bremer HJ, Herzog V, *et al.* A metabolic disorder similar to Zellweger syndrome with hepatic acatalasia and absence of peroxisomes, altered content and redox state of cytochromes, and infantile cirrhosis with hemosiderosis. *Europ J Pediatr* 1977; 124: 261-75.

18. Pagon RA, Graham JM Jr, Zonana J, Yong S-L. Coloboma, congenital heart disease, and choanal atresia with multiple anomalies: CHARGE association. *J Pediatr* 1981; 99:223-27.

GENETIC COUNSELING IN PSYCHIATRIC DISORDERS

John D. Rainer

There are few clinical areas in which problems in diagnosis affect genetic counseling more than for the major psychiatric disorders. Added to these diagnostic problems is the gap between hypotheses currently under investigation regarding modes of inheritance and any application to practical problems in counseling. Finally some role for the environment, including pre-natal, familial, and social aspects, is clear in all data, and yet its nature and magnitude are at least as poorly defined as the genetic components involved. These factors, known to the profession and suspected by the public, may account for the relatively low demand for genetic counseling in psychiatry. However, this picture is changing and those of us called upon to counsel patients, families, and couples regarding genetic aspects of mental illness must rely on the following equipment:

1. A conviction based on an extensive research literature that genes play an important role.
2. A finely tuned clinical diagnostic acumen based on first-hand experi-ence with patients and bolstered by a reliable official manual of criteria for classification.
3. A critical and selective use of published empirical risk figures and suggestions of modes of inheritance.
4. An up-to-date knowledge of prognosis, treatment, and biological factors.
5. An experience in individual and family dynamics permitting evaluation of psychological influences on disease expression, as well as the psycho-logical techniques in counseling which I discussed at this forum four years ago.[1]

Let me give you one person's current view of counseling, particularly in schizophrenia and the affective disorders, and some examples reflecting the complexities found in practice, taking up separately the five areas noted above.

CLINICAL GENETICS: PROBLEMS IN
DIAGNOSIS AND COUNSELING

187

EVIDENCE FOR GENETIC FACTORS

A half century of research has established to most people's satisfaction that genes have a role in the transmission and expression of the schizophrenic and affective disorders.

Early studies of risk in kinships of affected persons consistently showed significantly greater morbidity than in the general population for first degree relatives, with a lesser increase in second degree relatives. Most investigators acknowledged the difficulty in separating genetic from environmental influences in such familial transmission, and in fact more recent models tend to specify transmission of a vulnerability which is both genetically and environmentally determined.[2]

Separating these two influences in a quasi-experimental fashion depended on investigations of populations of twins and adoptees. In the one instance, concordance for illness in pairs with varied genetic relationships (MZ twin, DZ twin, sib, 1/2 sib, step sib) could be compared, with the home environment relatively constant. In the second instance, biological relationship was constant and family environment differed. Thus the relative influence of each could be measured. Studies of twins abounded in schizophrenia[3] and to a lesser extent in affective disorders. Adoption studies have been conducted in schizophrenia, psychopathy, and alcoholism,[4] and more recently have appeared in manic-depressive illness.[5] Neither of these approaches is without bias: MZ and DZ twin pairs in the same home might not have the same respective environment; adoptions may not be made at random so that biological and adoptive homes may have significant environmental correlation.

To be sure, some clinicians have been oversold by this body of data while others remain dubious. The former tend to ignore or discredit studies of social and rearing environment and life events, while the latter's critiques range from *ad hominem* and political attacks on the investigators to legitimate inquiries into research design, diagnosis, and statistical handling of data. In total, however, with the body of evidence, no responsible genetic counselor would maintain to a client that genes play no role in the transmission of psychotic disorders.

ADVANCES IN DIAGNOSIS

Diagnosis in the major psychoses has undergone enough of a change in the last five to ten years to make much previous research subject to reanalysis. For many years schizophrenia was over-diagnosed by current standards, at least in the United States, while the diagnosis of manic-depressive psychosis (not yet differentiated into bipolar and unipolar forms — depression with and without manic episodes respectively) was correspondingly less assigned. The diagnosis of schizophrenia included a wide spectrum of conditions ranging from acute, self-limiting psychotic episodes (with florid symptoms of elated mood and state of consciousness) to chronic withdrawal, from hallucinatory and delusion states to borderline and schizoid personality. The current diagnostic manual of the American Psychiatric Association (DSM III)[6] limits the diagnosis of schizophrenic disorder to one marked in the active phase by particular kinds of delusions or hallucinations or marked thought disorder, together with defined prodromal and residual phases, all lasting for at least six months. Acute, schizoid, and borderline symptoms are defined separately. If elevated and expansive or depressive mood with accompanying behavior is paramount, the diagnosis of major affective disorder is made. Thus an attempt is made to define heterogeneity at the clinical level, for which genetic (and biochemical) studies may provide further refinement.

EMPIRICAL RISK DATA

Our armamentarium of empirical risk data and prognostic indicators on which so much of genetic counseling for these disorders is based was obtained for many years as a by-product of attempts to prove a genetic role in these disorders and, more recently, to determine the mode of inheritance. As a result these data vary considerably according to the nature of the populations studied and the diagnostic systems being considered.

For schizophrenia the rates most pertinent for counseling are those for children of one or two schizophrenic parents and for nephews, nieces, and grandchildren. From the literature, it appears that the risk for children with one parent definitely affected is about 12%, with both parents affected about 37%, for nephews and nieces just over 2%, and for grandchildren about 3%. The latter figures for second degree relatives are particularly useful since it is often an unaffected sibling who comes for counseling. In the general population the risk is usually taken as almost 1%, though a number of studies have pointed to a higher figure.

The risk figures given represent a taking-off point and need to be tempered by the sophistication gained by familiarity with clinical data, consideration of many population studies, and with general genetic principles. For instance, there was indication as far back as Kallmann's Berlin family study in the 1930s[7] that paranoid schizophrenia posed a lesser risk for relatives than hebephrenic or catatonic forms. In the 1960s various studies of twins found a lesser risk for co-twins of individuals with milder forms of schizophrenia. At the same time, however, among the relatives of more severe schizophrenic patients milder forms and other psychiatric disorders of a schizoid nature were found in addition to the expected rate of the typical disorder. This finding is consistent with the paradigm of multifactor liability with one or more thresholds. Actually, subsequent studies of adoptees found little or no risk for relatives of patients with acute schizophrenia;[8] in fact there is strong opinion today that this episodic and florid self-limited psychosis is more clearly allied genetically to manic-depressive disease.

If the form of inheritance were indeed polygenic, the risk for a given descendant would depend also on the number of affected relatives, including whether they existed on both sides of the family; even if a major gene were involved, probably dominant, modifying factors might be polygenic and segregate separately. It can be seen in any event that among the requirements for counseling are an educated surmise about the genotype and phenotype of family members, and an assessment of organic factors.

In addition, an appreciation is necessary of developmental experiences that may play a role in determining illness in genetically vulnerable individuals. A crossfostering study seems to indicate that actual rearing by a schizophrenic parent does not seem to affect the risk for a child unless the child is biologically vulnerable.[9] What advice about the subtleties of child rearing the psychiatrist/counselor can give seems at present to consist of the ordinary rules of good family interaction and mental hygiene. Perhaps the longitudinal studies of high-risk children now underway[10] may provide means to spot the genetically vulnerable child, and suggest specific preventive measures to lower his or her risk. The discovery of specific linkage to genetic markers – at the DNA, enzyme or even trait level – would of course help immensely in such detection, though considerations of heterogeneity and environment would still prevail.

In the affective disorders many of the same principles noted apply; however the illness itself, though potentially destructive to life and career, is episodic and today largely controllable pharmacologically. The burden of illness, as differentiated from the risk, is thus lower.

Clinical heterogeneity in the affective disorders begins with the division into bipolar and unipolar disease: the former marked by periods of mania as well as depression, the latter by recurrent depression without any history of mania. The overall risk for unipolar depression in first degree relatives of unipolar patients has been given as 9-10% in males, 13-14% in females, as compared to 1.8% and 2.5% respectively for males and females in the general population. Of course reactive depression, depressive neurosis and minor depressive episodes have a much higher prevalence in the general population, up to 25%. Unipolar depression has been subdivided clinically by Winokur[11] into pure depressive disease which is found more often in males, has a late onset (after 40 years), with depression but not alcoholism in first degree relatives. So-called depressive spectrum disease is said to be more prevalent in women, has an early onset, and a higher risk in first degree relatives, with depression in female relatives, and alcoholism in males.

Bipolar disease is the less common in the population, but has the higher risk for first degree relatives, if unipolar disease when found in such relatives is considered a genetically equivalent condition. For bipolar illness, with a population estimate of about 1%, the risk for first degree relatives appears to run from 15-21%, with as high as 35% in some studies and over 50% for children, particularly daughters, in others. About half of the affected relatives of bipolar patients have depression alone, and females have been found in most studies to have a somewhat higher risk. If the affected family member had an early onset and a preponderance of manic symptoms, the genetic risk is greater, but importantly so is the treatment response to lithium. A survey of pedigrees suggests a dominant pattern of inheritance and this was borne out, at least in one series, by fitting risk figures to a single gene threshold model. There has been some excitement and controversy regarding the likelihood that an X-linked dominant gene close to the color blindness loci is responsible for some bipolar illness,[12] but the data are not specific enough for use in counseling. Finally, the development of a reliable biological test for affective psychosis holds promise; the usual suppression of serum cortisol levels by a single dose of dexamethasone tends not to occur, or to occur less completely, during endogenous depressive states.[13] This test can help to distinguish such biological, heritable disease from neurotic depression or loss and grief reactions. Its value in bipolar illness is under investigation.

In summary I would address those who say "If the diagnosis is in doubt, don't counsel " as follows:

1. It may be possible to clarify the diagnosis by looking first-hand at the individual case clinical data.
2. It may be necessary in selected cases to discuss differential diagnosis and respective outlook in each case.
3. If the facts are indeed poor, *e.g.* "nervous breakdown" with no more particulars available, specific counseling may indeed be withheld or postponed and the reason given.

It should be clear that genetic models can be used and even described to the more sophisticated client only as hypotheses for consideration along with discussion of empirical risk figures.

PROGNOSIS AND TREATMENT

Counseling in psychiatric illness with attention to the burden as well as the risk must include discussion of the prognosis, the range of severity, available methods of prevention and treatment in the condition under discussion. It has been said that psychiatric genetic counseling ought to be left to psychiatrists, who can appraise the status of the counselees, obtain and evaluate family histories, interpret and assess psychiatric reports.[14] Direct interview of family members (family study) has been shown to be more accurate than obtaining data from one member (family history).[15] Often records vary in completeness and sophistication. Here clinical experience and knowledge are imperative. Informed counseling may also make for better follow-up.

A recurrent problem in psychiatric genetic counseling is the risk to the fetus resulting from continuation of maternal psychotropic drug therapy during pregnancy, versus the risk of serious exacerbation of psychiatric illness if therapy is stopped. There are few definitive data on this score. Lithium is generally contraindicated,[16] while caution is suggested for other antidepressant and antipsychotic drugs. But no clear evidence is available as to their safety or hazard. Clearly there is a need for more definitive epidemiological studies. Meanwhile many decisions to avoid childbearing are being made by counselees on this ground rather than on genetic considerations.

PSYCHOLOGICAL INSIGHT AND CAPACITY FOR EMPATHY

Aside from the usual qualities and experience needed by any counselor in assessing the motives and fears of the client and helping the client to deal with the impact of genetic information, the psychiatric genetic counselor must be familiar with the effect of psychiatric disease on all members of the family, and conversely with the effect of family interaction on the manifestation of psychiatric disease. In the case of a family with one schizophrenic or manic parent, in addition to the genetic risk, it may be equally important to discuss:

1. The effect of having a child upon the parent's illness.
2. The effect of the parent's illness on the family stability and hence the psychological development of the child even aside from the child's genetic risk.

It has been asked whether psychiatric patients can use and understand genetic information without worsening their condition; this applies to many former psychiatric patients who have benefited from treatment. Again there is no substitute for clinical judgment, though reassurance as to genetic risk is actually more frequent than supposed.

It is clear that what we know and will learn about genetic factors in psychiatric illness in no way nullifies our knowledge about and concern with psychodynamic and psychological influences; both realms form part of a total biological approach and must converge in the art and science of psychiatric genetic counseling.

REFERENCES

1. Rainer JD. Psychiatric considerations in genetic counseling. *In*: Porter IH, Hook EB, eds. Service and education in medical genetics. New York: Academic Press, 1979:295-302.
2. Gottesman II, Shields J. A polygenic theory of schizophrenia. *Proc Nat Acad Sci* 1979; 58:199-205.
3. Gottesman II, Shields J. Schizophrenia and genetics. New York: Academic Press, 1972:24-37.
4. Crowe RR. Adoption studies in psychiatry. *Biol Psychiat* 1975; 10:353-371.

5. Mendlewicz J, Rainer JD. Adoption study supporting genetic transmission in manic-depressive illness. *Nature* 1977; 268:327-329.
6. American Psychiatric Association. Diagnostic and statistical manual of mental disorders (3rd ed.). Washington: APA, 1980.
7. Kallmann FJ. The genetics of schizophrenia. New York: JJ Augustin, 1938.
8. Kety SS, Rosenthal D, Wender PH, Schulsinger F, Jacobsen B. Mental illness in the biological and adoptive families of adopted individuals who have become schizophrenic. *In*: Fieve RR, Rosenthal D, Brill H, eds. Genetic research in psychiatry. Baltimore: Johns Hopkins Univ Press, 1975:147-166.
9. Wender PH, Rosenthal D, Kety SS, Schulsinger F, Welner J. Cross-fostering: a research strategy for clarifying the role of genetic and experiential factors in the etiology of schizophrenia. *Arch Gen Psychiat* 1974; 30:121-128.
10. Erlenmeyer-Kimling L, Cornblatt B, Fleiss J. High-risk research in schizophrenia. *Psychiat Annals* 1979; 9:79-102.
11. Winokur G, Cadoret RJ, Dorzab J. Depressive disease: a genetic study. *Arch Gen Psychiat* 1971; 24:135-144.
12. Mendlewicz J, Fleiss JL. Linkage studies with X chromosome markers in bipolar (manic-depressive) and unipolar (depressive) illness. *Biol Psychiat* 1974; 9:261-294.
13. Carrol BJ, Feinberg M, Greden JF, *et al.* A specific laboratory test for the diagnosis of melancholia. *Arch Gen Psychiat* 1981; 38:15-22.
14. Roberts JAF. An introduction to medical genetics. Oxford: Oxford Univ Press, 1967.
15. Mendlewicz J, Fleiss J, Cataldo M, Rainer JD. Accuracy of the family history method in affective illness. *Arch Gen Psychiat* 1975; 32:309-314.
16. Shepard TH. Catalog of teratogenic agents (3rd ed.). Baltimore: Johns Hopkins Press, 1980:199-200.

SCREENING AND PRENATAL DIAGNOSIS OF CYSTIC FIBROSIS: INTRODUCTION AND REVIEW

Kurt Hirschhorn

Since the reports in 1967 by Spock[1] on the presence of ciliary dyskinesia activity in sera from patients with cystic fibrosis (CF) and by Mangos on the presence in sweat[2] and saliva[3] from these patients of activity inhibiting sodium reabsorption, many experiments have been reported designed to diagnose the carrier state for CF. Recently, a series of studies have attempted to devise a method for the prenatal diagnosis of this disease. The high level of activity in the field has been spurred by the fact that CF is a debilitating, lethal and common hereditary disease. It is probably the most common genetic problem in whites, occurring once in about 1,600 to 2,500 live births, which translates into a carrier frequency of one in about 20 to 25, based on the assumption that all cases are due to inheritance of the identical original gene mutation. The validity of this assumption, *i.e.* the absence of genetic heterogeneity, has not been established and, in the light of the presence of heterogeneity in most other inborn errors, is probably unlikely.[4] If a significant amount of hetero- geneity does exist, leading incidentally to an even higher carrier frequency, much of the inconsistency in results relating to carrier detection and prenatal diagnosis could be based on such variation. In addition, since it is not clear that any of the proposed tests assay the primary gene product defective in CF, many of the experimental problems could be due to secondary alterations in the product being tested.

The earliest attempts at carrier detection were based on several systems involving disturbance or cessation of ciliary activity, using material from serum[1,5-7] fibroblasts[8,9] and lymphocytes.[10] The ciliated test materials have included rabbit trachea,[1,7,9,10] oysters[5,8] and mussels,[6] as well as flagellated bacteria.[11] The early workers in this field included Bowman and her associ- ates,[5] also collaborating with Danes and Bearn,[12] Conover and Beratis and their associates,[7,9,10] and Besley *et al.*[6] Much useful information derived from these early studies, including the demonstration that ciliary factor is a small molecule with a molecular weight of less than 10,000, loosely bound in serum to IgG,[13,14] cationic[5] with a significant amount of arginine, and possibly con- sisting of more than one moiety.[15] Ciliary dyskinesia activity was also found in sera, but not cells, from asthmatics and some patients with autoimmune dis-

CLINICAL GENETICS: PROBLEMS IN
DIAGNOSIS AND COUNSELING

orders.[16] At least one of the serum molecules was shown to resemble C3a,[13] an anaphylatoxin derived from the third component of complement, which may have been responsible for the false positives in other patients due to its elevation in several diseases including CF.[17] Of the early workers, only Bowman and her group have continued to pursue this line of investigation. She has been able to purify a glycoprotein(s) in urine of CF patients with ciliary dyskinesia activity.[18] Antibodies to this substance cross-react with some normal glycoprotein(s). She has begun attempts to map the gene[19] for this glycoprotein on the assumption, false in my opinion, that it represents the primary gene product responsible for CF. Based on these early findings, Wilson and Fudenberg have done a large number of studies,[20-23] primarily by iso-electric focusing, which have a reasonable likelihood of leading to further purification and characterization of the factor(s). Their methods have already led to the development by Manson and Brock[24] of an antiserum to the semipurified factor, which shows some promise as a reagent for carrier detection and prenatal diagnosis. Nelson's group[25] is studying a ciliostatic factor in saliva from patients with CF complexed with amylase, which inhibits a glycogen debranching enzyme. Their factor has a molecular weight of less than 1,000 and could be a breakdown product of one of the factors found in serum. The assay is difficult, and the results in heterozygotes overlap those in normal and abnormal homozygotes.

Based on a hypothesis[13,15,26] that the accumulation of arginine-rich peptides is due to a possible primary defect in an arginine esterase or peptidase, Nadler and his co-workers have developed separative and quantitative techniques which have shown that at least one such esterase activity may be diminished in the serum of patients.[27] More recently they have applied their techniques to the prenatal diagnosis of CF[28,29] and the results appear promising, although they may not have reached the 100% accuracy desirable for a prenatal test. Some work in their laboratory, particularly that done by Shapira[30,31] and Wilson and Fudenberg,[21] addressed the question of whether the abnormality in esterase activity was caused by a defect in the antiprotease properties of alpha$_2$-macroglobulin activity. While this approach seems to be dormant at this time, Shapira recently has begun to work directly on ciliary factors, both with a biochemical and an immunological approach. Blitzer and Shapira[32] have tentatively identified a ciliary factor in serum from patients with CF which again is a glycopeptide of approximately 5,000 molecular weight, and are attempting to produce monoclonal antibodies against this substance.

A few years ago there was a flurry of activity initiated by Epstein and Breslow, who reported that skin fibroblasts from patients with CF are relatively resistant to the toxic effects of steroids,[33] cyclic AMP[34] and cardiac glycosides.[35] They recently reported that such cells accumulate less sodium from the medium than normal cells.[36] While the latter claim has been withdrawn,[37] the first observation, derived from the early observations of Mangos[2] implying an abnormal pump mechanism, remains to be pursued. A variety of other studies have addressed the search for factors that inhibit sodium transport.

Another approach which has not been adequately pursued has been the description of a lectin in the blood of CF patients and carriers, which agglutinates red cells.[38] Also, elevated levels of polyamines have been described by Rennert and others[39] − another unconfirmed lead. Shapiro and his associates have concentrated on abnormalities of intracellular calcium,[40] mitochondrial NADH dehydrogenase[41] and fibroblast senescence[42] associated with reduced thymidine incorporation.[43] None of these approaches have been verified.

Some work deriving from the ciliary studies has led to interesting demonstrations of increased mucus secretion induced by serum from patients with CF[44] using a variety of assay systems. These findings would go along with the known hypersecretion of mucus in such patients.

Additional approaches have included the demonstration of a leucocyte degranulating factor in serum from patients with CF,[15] the study of metachromasia in fibroblasts from patients and carriers,[45] the latter being too nonspecific to be useful. The search for linkage between the CF locus and polymorphic genes, so far has not been successful. This last approach should become useful with the application of the recently discovered high level of polymorphism in many restriction endonuclease sites in human DNA, an approach already proven valuable in hemoglobinopathies.[46] This technique may lead to the identification, isolation and characterization of the CF gene.

A critical examination of the enormous amount of work described above makes me both depressed and optimistic. I am disappointed that the vast expense and effort have not led to the identification of the primary genetic defect in CF nor to a universally acceptable means of heterozygote detection or prenatal diagnosis. However, I am convinced that the work has produced enough clues to lead scientists involved in the field toward a resolution of these problems. It is important to emphasize the point made by Desnick at a previous Birth Defects Symposium[47] that availability of a test for carrier detection, accompanied by accurate prenatal diagnosis, will lead to new and difficult problems in the delivery of these services to the many families potentially at risk for having a child with CF.

REFERENCES

1. Spock A, Heick HMC, Cress H, Logan WS. Abnormal serum factor in patients with cystic fibrosis of the pancreas. *Pediat Res* 1967; 1: 173-77.
2. Mangos JA, McSherry NR. Sodium transport: Inhibitor factor in sweat of patients with cystic fibrosis. *Science* 1967; 158: 135-36.
3. Mangos JA, McSherry N, Benke P. A sodium inhibitory factor in the saliva of patients with cystic fibrosis of the pancreas. *Pediat Res* 1967; 1: 436-42.

4. Hirschhorn K. Genetic studies in disease. *In*: Mangos JA, Talamo RC, eds. Fundamental problems of cystic fibrosis and related diseases. New York: Intercontinental Medical Book Corp., 1973: 11-20.
5. Bowman BH, Lockhart LH, McCombs ML. Oyster ciliary inhibition by cystic fibrosis factor. *Science* 1969; 164: 325-26.
6. Besley GTN, Patrick AD, Norman AP. Inhibition of the motility of gill cilia of dreissenia by plasma of cystic fibrosis patients and their parents. *J Med Genet* 1969; 6: 278-80.
7. Conover JH, *et al*. Studies on ciliary dyskenesia factor in cystic fibrosis. I. Bioassay and heterozygote detection in serum. *Pediat Res* 1973; 7: 220-23.
8. Danes BS, Bearn AG. Oyster ciliary inhibition by cystic fibrosis culture media. *J Exp Med* 1972; 136: 1313-17.
9. Beratis NG, Conover JH, Conod EJ, Bonforte RJ, Hirschhorn K. Studies on ciliary dyskinesia factor in cystic fibrosis. III. Skin fibroblasts and cultured amniotic fluid cells. *Pediat Res* 1973; 7:958-64.
10. Conover JH, Beratis NG, Conod EJ, Ainbender E, Hirschhorn K. Studies on ciliary dyskinesia factor in cystic fibrosis. II. Short term leukocyte cultures and long term lymphoid lines. *Pediat Res* 1973; 7:224-28.
11. Cohen FL, Daniel WL. Effects of cystic fibrosis sera on proteus vulgaris motility. *J Med Genet* 1974; 11: 253-56.
12. Bowman BH, Barnett DR, Matalon R, Danes BS, Bearn AG. Cystic fibrosis: Fractionation of fibroblast media demonstrating ciliary inhibition. *Proc Nat Acad Sci USA* 1973; 70: 548-51.
13. Conover JH, Conod EJ, Hirschhorn K. Studies on ciliary dyskinesia factor in cystic fibrosis. IV. Its possible identification as anaphylatoxin (C3a) – IgG complex. *Life Sci* 1974; 14: 253-66.
14. Bowman BH, Lankford BJ, Fuller GM, Carson SD, Kurosky A, Barnett DR. Cystic fibrosis: The ciliary inhibitor is a small polypeptide associated with immunoglobulin G. *Biochem Biophys Res Comm* 1975; 64: 1310-15.
15. Conod EJ, Conover JH, Hirschhorn K. Demonstration of human leukocyte degranualtion induced by sera from homozygotes and heterozygotes for cystic fibrosis. *Pediat Res* 1975; 9: 724-29.
16. Conover JH, Conod EJ, Hirschhorn K. Ciliary dyskinesia factor in immunological and pulmonary disease. *Lancet* 1973; 1: 1194.
17. Conover JH, Conod EJ, Hirschhorn K. Complement components in cystic fibrosis. *Lancet* 1973; 2: 1501.
18. McNeely MC, *et al*. Cystic fibrosis: II. The urinary mucociliary inhibitor. *Pediat Res* 1981. (in press)
19. Mayo BJ, Klebe RJ, Barnett DR, Lankford BJ, Bowman BH. Somatic cell genetic studies of the cystic fibrosis mucociliary inhibitor. *Clin Genet* 1980; 18: 379-86.
20. Wilson GB, Fudenberg HH, Jahn TL. Studies on cystic fibrosis using isoelectric focusing. I. An assay for detection of cystic fibrosis homozygotes

and heterozygote carriers from serum. *Pediat Res* 1975; 9: 635-40.

21. Wilson GB, Fudenberg HH. Studies on cystic fibrosis using isoelectric focusing. II. Demonstration of deficient proteolytic cleavage of alpha$_2$-macroglobulin in cystic fibrosis plasma. *Pediat Res* 1976; 10: 87-96.

22. Wilson GB, Monsher MT, Fudenberg HH. Studies on cystic fibrosis using isoelectric focusing. III. Correlation between cystic fibrosis protein and ciliary dyskinesia activity in serum shown by a modified tracheal bioassay. *Pediat Res* 1977; 11: 143-46.

23. Wilson GB, Fudenberg HH. Studies on cystic fibrosis using isoelectric focusing. IV. Distinction between ciliary dyskinesia activity in cystic fibrosis and asthmatic sera and association of cystic fibrosis protein with the activity in cystic fibrosis serum. *Pediat Res* 1977; 11: 317-24.

24. Manson JC, Brock DJH. Development of a quantitative immunoassay for the cystic fibrosis gene. *Lancet* 1980; 1: 330-31.

25. Impero JE, Harrison GM, Nelson TE. Specificity of an isolated salivary factor material to cystic fibrosis. *Pediat Res* 1981; 15:940-944.

26. Nadler HL. Enzyme studies. *In*: Mangos JA, Talamo RC, eds. Cystic fibrosis: Projections into the future. New York: Stratton Intercontinental Medical Book Corp., 1976: 285-90.

27. Rao GJS, Walsh-Platt M, Nadler HL. Reaction of 4-methylumbelliferyl-guanidinobenzoate with proteases in plasma of patients with cystic fibrosis. *Enzyme* 1978; 23: 314-19.

28. Walsh MMJ, Nadler HL. Methylumbelliferylguanidinobenzoatere active proteases in human amniotic fluid. Promising marker for the intrauterine detection of cystic fibrosis. *Am J Obstet Gynecol* 1980; 137: 978-82.

29. Nadler HL, Walsh MMJ. Intrauterine detection of cystic fibrosis. *Pediatrics* 1980; 66: 690-92.

30. Shapira E, Ben-Yoseph Y, Nadler HL. Decreased formation of alpha$_2$-macroglobulin-protease complexes in plasma of patients with cystic fibrosis. *Biochem Biophys Res Comm* 1976; 71: 864-70.

31. Shapira E. Martin AL, Nadler HL. Comparison between purified alpha$_2$-macroglobulin preparations from normal controls and patients with cystic fibrosis. *J Biol Chem* 1977; 252: 7923-29.

32. Blitzer MG, Shapira E. A purified serum glycopeptide from controls and cystic fibrosis patients. I. Comparison of their mucociliary activity on rabbit tracheal explants. *Pediat Res*. (in press)

33. Breslow JL, Epstein J, Fontaine JH, Forbes GB. Enhanced dexamethasone resistance in cystic fibrosis cells: Potential use for heterozygote detection and prenatal diagnosis. *Science* 1978; 201: 180-82.

34. Epstein J, Breslow JL, Fitzsimmons MJ, Vayo MM. Pleiotropic drug resistance in cystic fibrosis fibroblasts: Increased resistance to cyclic AMP. *Somatic Cell Genet* 1978; 4:451-64.

35. Epstein J, Breslow JL. Increased resistance of cystic fibrosis fibroblasts to ouabain toxicity. *Proc Nat Acad Sci USA* 1977; 74: 1676-79.

36. Breslow JL, McPherson J, Epstein J. Distinguishing homozygous and heterozygous cystic fibrosis fibroblasts from normal cells by differences in sodium transport. *N Engl J Med* 1981; 304: 1-5.

37. Breslow JL, McPherson J. Sodium transport in cystic fibrosis fibroblasts not different from normal. *N Engl J Med* 1981; 305: 98.

38. Lieberman J, Kaneshiro W, Costea NV. Presence of a serum hemagglutinin (lectinlike factor) in cystic fibrosis homozygotes and heterozygotes. *J Lab Clin Med* 1981; 97: 646-53.

39. Rennert OM, Frias J, LaPointe D. Methylation of RNA and polyamine metabolism in cystic fibrosis. *In:* Mangos JA, Talamo RC, eds. Fundamental problems of cystic fibrosis and related diseases. New York: Intercontinental Medical Book Corp., 1973: 41-52.

40. Feigal RJ, Shapiro BL. Altered intracellular calcium in fibroblasts from patients with cystic fibrosis and heterozygotes. *Pediat Res* 1979; 13: 764-68.

41. Shapiro BL, Feigal RJ, Lam LF. Mitochondrial NADH dehydrogenase in cystic fibrosis. *Proc Nat Acad Sci USA* 1979; 76: 2979-83.

42. Shapiro BL, Lam LF, Fast LH. Premature senescence in cultured skin fibroblasts from subjects with cystic fibrosis. *Science* 1979; 203: 1251-53.

43. Lam LF, Shapiro BL. Differential incorporation of $_3$H-Thymidine into DNA in cultured skin fibroblasts derived from patients with cystic fibrosis and controls. *Life Sci* 1979; 24: 2483-90.

44. Czegledy-Nagy E, Sturgess JM. Cystic fibrosis: Effects of serum factors on mucus secretion. *Lab Invest* 1976; 35: 588-95.

45. Danes BS, Bearn AG. Cystic fibrosis of the pancreas, a study in cell culture. *J Exp Med* 1969; 129: 775-94.

46. Kan YW, Lee KY, Furbetta M, Angius A, Cao A. Polymorphism of DNA sequence in the beta-globin gene region: Application to prenatal diagnosis of beta^O thalassemia in Sardinia. *N Engl J Med* 1980; 302: 185-88.

47. Desnick RJ, Sklower SL. Cystic fibrosis: Prospects for prospective genetic screening. *In:* Porter IH, Hook EB, eds. Birth defects: Service and education in medical genetics. New York: Academic Press, 1979: 237-252.

ANTENATAL DETECTION OF CYSTIC FIBROSIS

Henry L. Nadler
Kathi Mesirow
Phyllis Rembelski

During the past decade, numerous attempts have been made to develop an accurate and reliable test for the intrauterine detection of cystic fibrosis (CF). The approaches taken to date have utilized either cell-free amniotic fluid or cultivated amniotic fluid cells. Utilizing cell-free amniotic fluid in an attempt to establish a technique for the antenatal diagnosis of CF, Bowman et al detected "CF mucociliary inhibitor" in a concentrated fraction of at least one amniotic fluid associated with a homozygous fetus.[1] Recently, Brock and Hayward,[2] and Dann and Blau[3] have confirmed our observation, discussed below, of the presence of methylumbelliferylguanidinobenzoate (MUGB) reactive proteases and arginine esterase in cell-free amniotic fluid.

A number of investigators have attempted to extend observations reported in skin fibroblasts cultivated from patients with CF to cultivated amniotic fluid cells. These include: (a) cellular metachromasia,[4] (b) CF mucociliary inhibitor (CFMI),[5,6] (c) response to Tamm-Horsfall glycoprotein,[7] (d) enhanced dexamethazone-resistance,[8] and (e) inhibition of oxygen consumption.[9] Each of these parameters has been controversial and, more importantly, no extensive experience documenting their validity, either prospective or retrospective for the intrauterine diagnosis of CF, has been reported by any of the investigators. Further refinements of these approaches may ultimately prove useful for the intrauterine detection of CF.

Several studies from our laboratory have shown that saliva and plasma of patients with CF is deficient in the proteolytichydrolysis of arginine esters.[10,11] Recently, MUGB, an active site titrant of many serine proteases, was utilized to demonstrate this deficiency in activated as well as catalytically inactive plasma.[12-15] Significant differences in the extent of protease-MUGB reactivity were found when plasma and fibroblasts of patients with CF, obligate heterozygotes and control samples were compared.[14-16] In these studies, correction for the non-specific hydrolysis of MUGB in crude systems was accomplished by carrying out the reaction in the presence and absence of benzamidine, an efficient competitive inhibitor of trypsin-like enzymes.

CLINICAL GENETICS: PROBLEMS IN
DIAGNOSIS AND COUNSELING

An arginine esterase activity similar to that observed in plasma has been demonstrated in human amniotic fluid during the second trimester and at term. Like plasma, the protease(s) hydrolyzed esters of arginine were reactive towards MUGB and had a pI of 5.1–5.4. The pH optimum for proteolytic activity was 8.0. This protease activity was inhibited by soybean trypsin inhibitor (STI), benzamidine and (p-nitrophenyl)-p '-guanidinobenzoate (NPGB). Upon gel filtration, two MUGB-reactive fractions were observed, one with an apparent molecular weight of 200,000 and the other 100,000.

Both fractions had arginine esterase activity and appeared to be sensitive to inhibition by STI and benzamidine,[17,18] a finding confirmed by two other investigative groups.[2,3]

These observations have been extended retrospectively and, more recently,[19] prospectively to amniotic fluid from pregnancies at risk for CF.

MATERIALS AND METHODS

Amniotic fluid was obtained after informed consent from 79 pregnancies with a 25% risk of having a fetus with CF. In each case, the parents had previously given birth to a child with documented CF. Transabdominal amniocentesis was carried out between the 16th and 19th week of pregnancy, and the amniotic fluid sample, usually 10cc, was frozen and forwarded to our laboratory.

MUGB activity determination, isoelectric focusing on polyacrylamide gels and gel filtration was carried out as previously reported.[20,21] The criteria for the detection of CF included MUGB activity less than 1.35 nmoles MUGB per miligram of protein, a missing band on polyacrylamide electrophoresis and an abnormal gel filration pattern. The diagnoses were confirmed utilizing at least one and usually two sweat tests.

RESULTS

Based on the criteria listed above, 66 pregnanceis were predicted to be normal and 13 affected (Table 1). To date, 43 cases have been evaluated. Twenty-nine were predicted to be normal and were confirmed as being normal. Five were predicted to have CF and confirmed after delivery as being affected. Six were predicted to have CF and the pregnancies terminated. Two and probably a third case predicted to be normal did in fact have CF.

The mean MUGB activity in high risk and control amniotic fluids is shown in Table 1. As can be seen, the predicted normals had an activity lower than the standard controls. This is not unexpected since two-thirds of the nonaffected would presumably be carriers. The difference between the predicted population with CF, excluding the discrepant and missed cases, is significantly less than the control population, P<.001.

TABLE 1
MUGB Activity in High-Risk and Control Amniotic Fluids

	Number	nmoles MUGB/ mg protein
Controls	2000	2.38
Predicted normals (25% risk group)	5/6	1.94
Predicted CF (25% risk group) (excludes 2 discrepant and missed cases)	11	1.20

Three cases of CF, two confirmed, have been missed. In each of these, the specific activity was normal as were the banding patterns and column filtration. The reason a third case is stated to be possibly incorrect is that, although the child had meconium ileus, the sweat tests have not been completed. Four cases in which there were discrepancies from the criteria outlined above were observed. In one case the specific activity was 1.32, four bands were seen on isoelectric focusing and a normal column was found. The prediction was of a normal fetus and after delivery the baby was confirmed to be normal. In two cases, the specific activity was 1.76 and 2.2; isoelectric focusing revealed only three BAEE bands, two MUGB bands and column filtration patterns were abnormal. These pregnancies, despite their "normal" specific activity were predicted to have CF and the pregnancies were electively terminated. The fourth case was of interest in that the specific activity could not be measured conclusively. This is the only case of this kind. Only three BAEE bands and two MUGB bands were found. The column was also abnormal. This case was predicted to be affected and the pregnancy is still in progress. Similar results were obtained in another laboratory carrying out studies on a divided sample from this patient.

DISCUSSION

This study indicates the potential value of monitoring high risk CF pregnancies with the quantitative and qualitative measurements of MUGB as previously described in our laboratory. There are no good explanations for the false negatives, i.e., missed cases. Whether these cases are examples of genetic heterogeneity is unknown as the basic biochemical defect is yet to be identified. Not listed in this series are the studies carried out on lower risk pregnancies, i.e., <0.5–2%. In these studies all cases predicted to be normal have been normal and three cases predicted to have CF were shown to have CF. To date there have been no false positives.

Further studies are needed to confirm and identify the variables which would possibly permit identification of all cases. One of the cases missed has been studied retrospectively with a new modification of the column technique, using sephacryl, and the results would suggest that that fetus showed positive effect. The ability to confirm the diagnosis of CF in a presumably affected fetus would be of significant importance. Unfortunately, distinctive lesions have not been described in 16-22 week fetuses with CF, but the fibroblast techniques of Shapiro,[9] and Holsi,[7] if reproducible, may enable one to confirm the diagnosis.

We are presently offering to monitor, on a research basis, pregnancies at risk for CF. We are studying 2–3 cases per week. Within the next year the series should be large enough to validate the approach described in this report. Fortunately, a number of other laboratories around the world are also pursuing this approach.

ACKNOWLEDGMENT

This project was supported by grants from the National Foundation, March of Dimes and National Institutes of Health.

REFERENCES

1. Bowman BH, Lankford BJ, McNeely MC, Carson SD, Barnett DR, Berg K. Cystic fibrosis: Studies with the oyster ciliary assay. *Clin Genet* 1977; 12: 333-43.
2. Brock DJH, Hayward C. Methylumbelliferylguanidinobenzoate reactive proteases and prenatal diagnosis of cystic fibrosis. *Lancet* 1979; i: 1244.
3. Dann LG, Blau K. Arginine esterase in amniotic fluid: Possible marker for cystic fibrosis. *Lancet* 1979; ii: 907.
4. Danes BS, Bearn AG. Hurler's syndrome. A genetic study in cell culture. *J Exp Med* 1966; 123: 1-16.
5. Beratis NG, Conover JH, Conod EJ, Bonforte RJ, Hirschhorn K. Studies of ciliary dyskinesia factor in cystic fibrosis. III. Skin fibroblasts and cultured amniotic fluid cells. *Pediatr Res* 1973; 7: 958-64.
6. Bowman BH, Lockhart LH, Herzberg VL, Barnett DR, Kramer J. Cystic fibrosis: Synthesis of ciliary inhibitor by amniotic cells. *J Clin Genet* 1973; 4: 461-63.
7. Hosli P, Vogt E: Reliable detection of cystic fibrosis in skin-derived fibroblast cultures. *Hum Genet* 1978; 41: 169.
8. Breslow JL, Epstein J, Fontaine JH, Forbes GB. Enhanced dexamethasone resistance in cystic fibrosis cells: Potential use for heterozygote detection and prenatal diagnosis. *Science* 1978; 201: 180-82.
9. Shapiro BL, Feigal RJ, Lam LFH. Mitochondrial NADH dehydrogenase in cystic fibrosis. *Proc Natl Acad Sci USA* 1979; 76: 2979-83.

10. Rao GHS, Nadler HL. Deficiency of trypsin-like activity in saliva of patients with cystic fibrosis. *J Ped* 1972; 80: 573-76.
11. Chan KYH, Applegarth DA, Davidson AGF. Plasma arginine esterase activity in cystic fibrosis. *Clin Chim Acta* 1977; 74: 71-5.
12. Jameson GW, Roberts DW, Adams RW, Hyle WSA, Elmore DT. Determination of the operational molarity of solutions of bovine alpha-chymotrypsin, trypsin, thrombin and factor Xa by spectroflurimetric titration. *Biochem J* 1973; 131: 107.
13. Kerr MA, Walsh DA, Neurath H. Catalysis by serine proteases and their zymogens. A study of acyl intermediates by circular dichroism. *Biochem* 1975; 14: 5088.
14. Rao GJS, Walsh-Platt M, Nadler HL. Reaction of 4-methylumbelliferylguanidinobenzoate with proteases in plasma of patients with cystic fibrosis. *Enzyme* 1978; 23: 314-19.
15. Walsh-Platt MM, Rao GJS, Nadler HL. Protease deficiency in plasma of patients with cystic fibrosis: Reduced reaction of 4-methylumbelliferylguanidinobenzoate with plasma of patients with cystic fibrosis. *Enzyme* 1979; 24: 224-29.
16. Walsh-Platt MM, Rao GJS, Nadler HL. Reaction of 4-methylumbelliferylguanidinobenzoate with cultivated skin fibroblasts derived from patients with cystic fibrosis. *Pediatr Res* 1978; 12: 874-77.
17. Walsh-Platt MM, Nadler HL. MUGB-reactive proteases in amniotic fluid: Possible marker for the intrauterine diagnosis of cystic fibrosis. *Lancet* 1979; i: 622.
18. Walsh MMJ, Rao GJS, Nadler HL. Reaction of 4-methylumbelliferylguanidinobenzoate with proteases in amniotic fluid. *Pediatr Res* 1980; 14: 353-56.
19. Walsh MMJ, Nadler HL. Methylumbelliferylguanidinobenzoate - reactive proteases in human amniotic fluid: Promising marker for the intrauterine detection of cystic fibrosis. *Am J Obstet Gynecol* 1980; 137: 978-82.
20. Nadler HL, Walsh MMJ. Intrauterine detection of cystic fibrosis. *Pediatrics* 1980; 66: 690-92.
21. Nadler HL. Prenatal detection of cystic fibrosis. *In*: Galjaard H. ed. The Future of Prenatal Diagnosis. Edinburgh: Churchill Livingston, 1981. In press.

ADDENDUM

More recent experience has shown a number of false positives, *i.e.* CF predicted, normal children delivered. This is a serious problem which if not overcome limits the usefulness of the present approach to the intrauterine diagnosis of CF.

PRENATAL SCREENING FOR CYSTIC FIBROSIS

David J. H. Brock

INTRODUCTION

The idea of prenatal screening for cystic fibrosis (CF) of the pancreas has been stimulated by successes in screening programs for two other recessively-inherited genetic diseases, namely Tay-Sachs and β-thalassemia. It is also influenced by the enormous experience which has been gained in the United Kingdom in early maternal screening for a group of quasi-genetic disorders, the open neural tube defects. The published results of these prenatal screening programs have shown that they can have a major impact on the birth prevalence of children with congenital defects.

It may well seem that prenatal testing is a somewhat draconian approach to take to a disorder like CF, where the median survival time has improved so much in the past twenty years, and where many affected infants born today may expect to live into and through their second decade. However, it must be noted that the improved prognosis in CF is a two-edged sword, and that longer survival can impose greater rather than lesser burdens on patients and their families. There is no doubt that recent encouraging advances in prenatal diagnosis of CF have been seized on enthusiastically by families with a previous affected child. Given the opportunity of a monitored pregnancy such parents will go to considerable lengths to avoid the disaster of a second affected child. The principle of prenatal screening assumes that such a response will also be applicable to a first affected child, if and when the technology for prevention becomes available. In parallel with most other simple diseases, prenatal screening for CF represents at present the only feasible approach to a reduction in the birth incidence of this distressing condition.

If large scale prenatal testing for CF is to become possible two requirements must be met. The first is a precise and reliable form of prenatal diagnosis in high-risk pregnancies. This subject is dealt with in detail by Nadler in another chapter. Though the methylumbelliferylguanidinobenzoate (MUGB) protease system currently in use has several imperfections, its achievements have been impressive. It is probably not too optimistic to suggest that refinements in MUGB protease analysis and measurement may soon provide a system with the type of reliability and precision needed for accurate diagnosis and thus for making decisions about whether or not a pregnancy might be terminated.

The second requirement in such testing is a method of heterozygote detection, so that at-risk couples may be identified and offered the option of prenatal diagnosis. If heterozygote detection is to be used in this way, above all it must be capable of handling very large numbers of samples. This effectively rules out bioassays, such as those used in the identification of the CF factors. It also rules out the use of skin fibroblasts or any type of complex biochemistry. In fact it is possible to predict that a suitable large-volume heterozygote detection system would have to be based on the analysis of serum samples by either immunological methods or automatable enzymatic systems. These requirements form the conceptual background to the work discussed in this paper.

AN IMMUNOLOGICAL APPROACH TO HETEROZYGOTE DETECTION

Some years ago Wilson and colleagues[1] observed the presence of a characteristic protein doublet at a pI of 8.4 when carefully collected serum samples from patients with CF (homozygotes) or carriers of the CF gene (heterozygotes) were subjected to high-resolution isoelectric focusing under denaturing conditions. The doublet responsible for the pI 8.4 band was named "CF protein" (CFP). The existence of CFP has been the subject of considerable controversy, mainly because of the difficulty of reproducing the stringent experimental conditions laid down by Wilson. It now seems certain, however, that CFP exists and is present in the serum of a majority of those who carry the CF gene in single or double dose.[2] Wilson and Fudenberg[3] have proposed isoelectric focusing as a general screening method; their argument is that since CF homozygotes will be diagnosed clinically and by sweat test, other individuals showing the pI 8.4 doublet can be labelled as heterozygotes. This assumes the diagnosis of CF homozygotes by classical methods to be of much greater precision than experience has recorded. Even when all the technical problems of isoelectric focusing are overcome, it tends to fail in discrimination in those frequent situations where the sibling of an affected proband may be either a mildly affected homozygote or a clinically unaffected heterozygote.

Some years ago we conducted trials of Wilson's isoelectric focusing system (work performed by Maurice Super and Caroline Hayward). A panel of three observers were trained in the interpretation of isoelectric focusing gels where the clinical status of the samples was known. Once they had achieved expertise and confidence on known gels, they were asked to perform blind reading of a series of unknown samples. A score of two was given to any sample in which the pI 8.4 doublet was unambiguously present, a score of one to a doubtful sample, and a score of zero to any sample where the doublet could not be seen. Samples from a panel of 23 CF homozygotes, 22 heterozygotes and 16

TABLE 1
Serum Isoelectric Focusing in the Scoring of CF Heterozygote
and Control Groups

Group	No.	Group mean ± SD	No. (%) misclassified
CF	23	1.43 ± 0.50	4 (17)
Heterozygote	22	1.22 ± 0.44	6 (27)
Control	16	0.45 ± 0.48	3 (19)

TABLE 2
Rocket Immunoelectrophoresis of Serum Samples
Against Guinea Pig Antiserum

Serum Samples	Positive Reaction	Negative Reaction
CF patients (56)	54	2 (3.7%)
Parents of CF patients (38)	35	3 (8.6%)
Apparent normals (49)	4 (8.2%)	45

apparently normal controls were each run twice and scored by the three observers. A sample which achieved an average score of greater than one was regarded as having a positive pI 8.4 doublet, and a sample with a score of less than one was regarded as negative. The proportion of subjects in each of three categories who were misclassified was unacceptably high (Table 1) leading to the conclusion that isoelectric focusing is an inadequate tool for the precise discrimination of CF homozygotes, heterozygotes and normals.[4] However, the mean scores for the affected homozygotes and heterozygotes were significantly greater than the mean score for the controls. This seemed to us to provide a clear indication of the existence of a protein entity similar to that described by Wilson as CFP.

Most attempts to improve the measurement of a protein of unknown function and properties have resorted to immunological methods. Our own procedure (work performed by Jean Manson) was to excise segments of the polyacrylamide gel around a pI 8.4 and to inject these homogenized fragments into guinea pigs (Figure 1). In this way an antiserum was raised, which after absorption with serum from a pI 8.4 negative individual, appeared directed

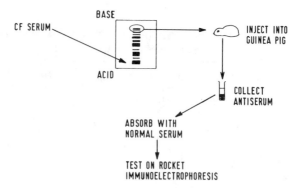

Fig. 1. Dann-Blau hypothesis (1978). Protocol for preparation of
antiserum against CFP.

against individuals who carried a single or double dose of the CF gene. Furthermore, the antiserum appeared capable of making a semi-quantitative distinction
between CF homozygotes and heterozygotes.[5]

Examination of the performance of the antiserum against a panel of CF
homozygotes, heterozygotes and normal controls produced a reasonable resolution of the three categories (Table 2). The misclassification of some of the
CF homozygotes and heterozygotes could be explained on the basis of the inevitable misdiagnosis of some individuals whose clinical signs suggested CF. It
could also be anticipated that some 4–5% of normal controls would be unrecognized heterozygotes. The observed proportions are, however, somewhat
higher than this, although the overall numbers are small (Table 2).

PROBLEMS IN THE IMMUNOASSAY OF CF PROTEIN

Attempts to develop the immunoassay against CFP have followed three
parallel courses: (a) the preparation of antisera in larger and more convenient
animals than the guinea pig, (b) the isolation of CFP by immunochromatography, and (c) a search for the functional nature of CFP.

Despite many attempts no specific antisera have been produced in larger
animals. High-titre antibodies have been prepared in sheep, goats and rabbits,
but after absorption these failed to react specifically with appropriate samples.
The reasons for this failure are unclear but may be related to the ratio of antigen in the excised acrylamide slices to the body mass of the larger animals.
Wilson has been successful in preparing an antiserum to CFP in mice,[6] which
suggests that there is nothing unique about the guinea pig. However, this has
not helped the problem of antiserum accumulation.

In our hands immunochromatography has so far failed to yield any product identifiable as CFP. It is possible that the avidity of the guinea pig anti-

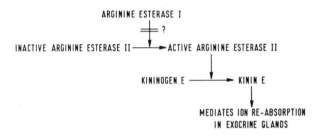

Fig. 2. Protocol for CF-specific antiserum.
Outline of the Dann-Blau hypothesis.

serum is too high to allow for the subsequent release of absorbed antigen. Alternatively, CFP may not be sufficiently stable to withstand the chaotropic ions used in stripping antiserum-linked columns.

Our third approach has been to search the literature for a possible clue to the identity of CFP. Here we have been attracted by the hypothesis recently proposed by Dann and Blau.[7] They suggested that the primary physiological defect in CF was the absence of an ion-mediating kinin, responsible for the control of reabsorption of sodium and chloride in exocrine glands. This kinin was assumed to be generated from a precursor kininogen by an arginine esterase-type proteolytic cascade (Figure2). Dann and Blau suggested that kininogen would accumulate in exocrine glands and serum of patients with CF and proposed that it might be either CFP or one of the well-known CF dyskinesis factors. Though the "kininogen-kinin" theory was based largely on circumstantial evidence, it was also supported by the reaction of our antiserum with salivary extracts from CF patients and the abolition of the reaction by prior treatment of extracts with commercial kallikreins.

AN ENZYMATIC APPROACH TO HETEROZYGOTE DETECTION

The reaction of our antiserum with samples from CF homozygotes and heterozygotes, and the absence of reaction with normal control samples, suggested to us that we were observing at best a secondary phenomenon in the pathology of the disorder. The Dann and Blau hypothesis, together with recent observations on the important role of MUGB proteases in prenatal diagnosis of affected fetuses, pointed to the possibility of an arginine esterase-type activity as the primary product of the mutant gene. We have, therefore, turned our attention to a scrutiny of a series of synthetic substrates, designed to

D-VAL-LEU-ARG-pNA

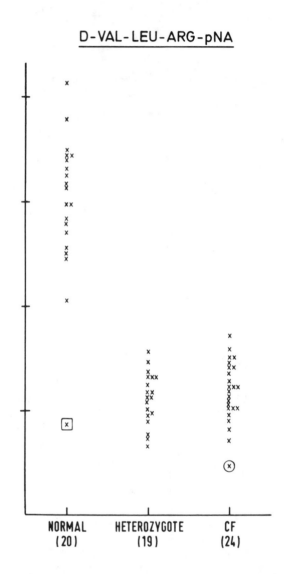

Fig. 3. Assay of serum samples using S2266 as substrate. ⊠ represents apparent heterozygote, ⊗ an adult cystic.

reveal arginine esterase activities. Of these the most effective have been a series of tripeptides, in which a terminal arginine is linked to a suitable chromophore.[8] Examination of a limited panel of CF and normal sera demonstrated that maximal discrimination could be achieved when D-valine was linked through leucine to the arginyl chromophore.

More detailed investigation of this substrate, marketed by Kabi Diagnostics under the trade name S2266, has given promising results. Assay of serum samples gave almost complete discrimination between normals on the one hand and CF homozygotes and heterozygotes on the other hand (Figure 3). The only serum sample from a normal control which fell within the range of individuals with the CF gene also reacted strongly with the guinea pig antiserum and is presumed to be an unrecognized heterozygote. However, the range of values obtained for heterozygotes and CF homozygotes overlapped completely. This somewhat puzzling result has recently been explained by the fact that all the heterozygotes and normals in the panel were adults, while all but one of the CF homozygotes were young children. The single exception was an adult patient with CF whose enzymatic value in this assay fell below the limits of the heterozygote range. A small number of other adults with CF have subsequently been tested and found to display values below the heterozygote range.

These preliminary results on the assay of an arginine esterase-type activity in serum with S2266 as substrate suggest that it should be possible ultimately to resolve completely the categories of normal, heterozygote and affected homozygote. Even if occasional overlap between heterozygotes and patients continues, the discrimination between normals and heterozygotes should make the method appropriate for large-scale screening, where the occurrence of a CF patient should be a rare event. The enzymatic assay is simple, cheap and suitable for automation. The commerical availability of the substrate means that it is possible to survey large numbers of serum samples, and to gain reliable information on the sensitivity and specificity of this postulated screening method.

ACKNOWLEDGMENTS

The work described here was supported by a generous grant from the Cystic Fibrosis Research Trust. We are grateful to Dr. Karl Blau of Queen Charlotte's Hospital, London, for first drawing our attention to the potentialities of S2266.

REFERENCES

1. Wilson GB, Jahn TL, Fonseca JR. Demonstration of serum protein differences in cystic fibrosis by isoelectric focusing in thin layer polyacrylamide gels. *Clin Chim Acta* 1973; 49: 79-84.

2. Nevin GB, Nevin NC, Redmond AO, Young IR, Tulley GW. Detection of cystic fibrosis homozygotes and heterozygotes by serum isoelectric focusing. *Hum Genet* 1981; 56: 387-89.
3. Wilson GB, Fudenberg HH. Is cystic fibrosis protein a diagnostic marker for individuals who harbor the defective gene? *Pediat Res* 1978; 12: 801-804.
4. Brock DJH, Hayward C, Super M. A blind trial of isoelectric focusing in detection of the cystic fibrosis gene. *Hum Genet* In press.
5. Manson JC, Brock DJH. Development of a quantitative immunoassay for the cystic fibrosis gene. *Lancet* 1980; 1: 330-32.
6. Wilson GB. Monospecific antisers, hybridoma antibodies and immunoassays for cystic fibrosis protein. *Lancet* 1980; 2: 313-14.
7. Dann L, Blau K. Exocrine-gland function and the basic biochemical defect in cystic fibrosis. *Lancet* 1978; 2: 405-07.
8. Claeson G, Aurell L, Karlsson G, et al. Design of chromogenic substrates. *In:* Scully MF, Kakkar VV, eds. Chromogenic peptide substrates: chemistry and clinical usage. Edinburgh:Churchill Livingstone, 1974: 20-31.

CYSTIC FIBROSIS: IMMUNOASSAYS FOR CARRIER DETECTION
AND METABOLIC CORRECTION *IN VITRO*

Gregory B. Wilson

INTRODUCTION

In the 1930s cystic fibrosis (CF) was first described as a specific disease entity by Fanconi et al[1] and Andersen.[2] Since then a great deal of effort has been devoted to identifying biochemical abnormalities unique to CF for the purpose of heterozygote detection and prenatal diagnosis.[3-10] While the major clinical manifestations including pancreatic insufficiency, elevated sweat electrolytes, and chronic pulmonary disease (the major cause of morbidity and mortality in CF patients) are believed to be secondary to exocrine gland dysfunction, the basic metabolic defect remains unknown. Up to now, few useful biochemical markers have been demonstrated for detecting heterozygote carriers.[3-5] One promising marker for CF genotypes, now termed the cystic fibrosis protein (CFP), was first demonstrated in serum in 1973[11] by analytical isoelectric focusing in thin-layer polyacrylamide gels (IEF-TLPG).

CFP occurs as one member of a band doublet in the uppermost centimeter (isoelectric point, pI 8.5) of the electrophoretogram resulting from IEF-TLPG of sera from CF genotypes (Figure 1). Several properties of CFP have been described previously (Table 1), and the diagnostic value of CFP has been confirmed by several other investigators using the IEF-TLPG method. Table 2 summarizes the frequency of CFP positive samples from CF genotypes and normal healthy controls. In addition, our results from sera from non-CF patient controls (Table 3) indicate that CFP is a specific marker for the CF gene.[12] Other than the interesting exception of non-CF patients with certain forms of leukemia (four out of five are CFP positive), the number of CFP positive samples is within the expected frequency of heterozygote carriers in the Caucasian population (approximately 5%). The finding of CFP in leukemic sera may suggest that CFP accumulates in the body fluids of CF genotypes due to a metabolic defect involving leukocytes and suggests that leukocytes (primarily monocytes [MNC]) may secrete CFP or a precursor of CFP.

CFP appears to be a good diagnostic marker for CF genotypes. However, analytic IEF-TLPG is technically difficult to perform and is not quantitative enough to distinguish between homo- and heterozygotes. Recently, mono-

TABLE 1

Summary of Properties of CFP

Property	Comments	Refs.
Isoelectric point 8.5	CFP is a negatively charged serum component which contains a polypeptide segment.	12,13
Purified with IgG by DEAE-cellulose chromatography at pH 8.0 or pH 8.6.	Confirms that CFP is negatively charged may also indicate that CFP is complexed to IgG.	14
Binds to Protein-A from *Staphlococcus aureus* covalently linked to Sepharose CL-4B	May indicate that CFP is complexed to IgG or CFP may bind directly to Protein A or to Sepharose. It is known that some serum components such as complement components C3 and C5 may bind to Sepharose.[16]	15
M.W. 3,500-10,000 determined by gel filtration chromatography, amicon ultrafiltration, and dialysis after treatment of serum or fractions obtained from serum with chaotropic agents (urea) or by acidification (pH 3.7).	Without prior acidification of the CFP containing sample, CFP is found in a serum fraction of M.W. between 90,000-180,000. CFP may be a fragment of a larger macroglobulin or a small molecule complexed to a carrier (IgG?) which is released at acid pH.	14,17
Found in sera from CF homozygotes and heterozygote carriers.	Directly or indirectly related to the primary genetic defect in CF.	11,12

TABLE 2
Reproducibility of the IEF–TLPG Method for Detecting CFP in Serum

Investigators	Serum	No.	CFP present* +	–	Freq. of CFP pos. (%)
Wilson et al[17]	Homo.	85	82	3	96.5
	Hetero.	66	62	4	93.9
	Nl. cont.	110	9	101	8.2
Altland et al[18]**	Homo.	3	3	0	100
	Hetero.	1	1	0	100
	Nl. cont.	3	0	3	0
Scholey et al[19]	Homo.	10	7,6	3,4	60,70
	Nl. cont.	16	0,0	16,16	0,0
Tulley et al[20]	Homo.	11	10	1	90.9
	Hetero.	9	8,7	1,2	88.9,77.8
	Nl. cont.	26	2	24	7.7
Manson & Brock[21]	Homo.	17	16	1	94.1
	Hetero.	9	8	1	88.9
	Nl. cont.	15	1	14	6.0
Nevin et al[22]	Homo.	20	18,18	2,2	90
	Hetero.	23	18,19	5,4	78.3,82.6
	Nl. cont.	100	8	92,92	8.0

*In most of the reports, samples were analyzed using a serum volume containing 300 μg IgG.[12] Where two values are noted they represent the results obtained when electrophoretograms were scored by two independent observers.
**The results noted by these investigators were obtained using a two step method consisting of electrofocusing in the first step and slab gel electrophoresis in the second step.

LIST OF ABBREVIATIONS

A_1AT	Alpha$_1$-antitrypsin
A_2M	Alpha$_2$-macroglobulin
BSA	Bovine serum albumin
CDS	Ciliary dyskinesia substances
CF	Cystic fibrosis
CFP	Cystic fibrosis protein
CIE	Counterimmunoelectrophoresis
DEAE	Diethylaminoethyl
EACA	Epsilon-amino-caproic-acid
IEF-TLPG	Isoelectric focusing in thin-layer polyacrylamide gels
IEP	Immunoelectrophoresis
M-Mϕ	Monocyte-macrophages
MNC	Mononuclear cells
M.W.	Molecular weight
PBS	Phosphate buffered saline
PMN	Polymorphonuclear, Neutrophils
RIEP	Rocket immunoelectrophoresis
SBTI	Soybean trypsin inhibitor
S.D.	Standard deviation

specific antisera to CFP raised in mice[23] and guinea pigs have been developed.[21] This report details our development of immunoassays for CFP using mouse antisera, our progress toward determining the identity of CFP, and the evidence indicating that CFP may be structurally related to one of the CF ciliary dyskinesia substances.

The ciliary dyskinesia substances (CDS) are a group of at least four polypeptide mediators[24,27] found in sera and produced by cultured cells from both CF genotypes (Table 4). The term CDS is derived from the production of marked disruption of normal ciliary movement (dyskinesia) in mammalian respiratory mucosal epithelial explants.[28] The more recent findings that the CDS can produce mucus expulsion, lysosomal degranulation, and alterations in electrolyte transport (K^+, Na^+, Ca^{++}) suggests that the CDS may be initiators of much of the exocrine gland dysfunction believed to underlie the pathogenesis of CF[24,33] Evidence is particularly strong for the involvement of those CDS (and other mediators[25,27]) produced by monocyte-macrophages (M-Mϕ) and T lymphocytes from CF genotypes in both the initiation of pulmonary obstruction and the accelerated progression of the destructive phase of pulmonary disease in CF, namely chronic inflammation.[24] The CDS (or other mediators produced by CF cells[6,24]) may be directly or indirectly

Fig. 1. Typical electrophoretograms obtained by isoelectric focusing (IEF) of sera from CFP positive CF genotypes 1,4 and CFP negative normal healthy controls or non-CF patient controls 2,3. IEF was performed in a pH 2.5-10 gradient.[12,13] The anode is at the bottom. The CFP band doublet is indicated by arrows; the lower band is CFP (pI 8.5).

TABLE 3

Frequency of CFP Detected by IEF in Other Disease States

Disease	Age (in years)	No. with CFP	No. Tested
Chronic allergy	10.6±6* (4–20)	1	8
Bronchial asthma	10.8±5.2 (3.5–28)	0	6
Bronchitis	5,7.6	0	2
Cirrhosis of the liver	37.4±20.3	1±	5
Diabetes mellitus	46,37,22	1	3
Hypokalemia with renal failure and hepatitis	54	1	1
Hypogammaglobulinemia	2,45	0	2
Kidney disease	44±14.8	0	9
Lupus erythematosus	40,59	0	2
Myocardial infarction	35,57	0	2
Other circulatory disorders	65.6±12.6	0	5
Pancreatitis	36	0	1
Recurrent pneumonia	52,3.5,8	0	3
Rhuematoid arthritis	52,45	0	2
Leukemia**	45±28 (6–74)	3+,1±	5
Bladder cancer	56	1	1
Multiple myeloma	40,65	0	2
Pancreatic cancer	51,55,58,40	1	4
Lung cancer	40,59	0	2
Hepatoma	22	0	1
Porphyria cutanea tarda	60	0	1
Cancer of the thalamus	49	0	1
Metastatic cancer of the bone	50	0	1
Total with CFP/Total studied		7+,2±	69

*Mean ± S.D., and range are listed when the number studied was five or more. Otherwise, all values are listed.

**3+, two acute granulocytic leukemia, one chronic monocytic leukemia; 1±, acute lymphocytic leukemia; 1–, acute lymphocytic leukemia. Modified from reference 12.

TABLE 4
Cells and Body Fluids of CF Genotypes Shown to Contain CDS*

Investigators	Ref.	Material
Spock et al, '67	28	
Conover et al, '73	29	Serum
Wilson et al, '77	14	
Conover et al, '73	30	Peripheral blood leukocyte cultures
Wilson et al, '78, '80	26,31	
Wilson et al, '78, '80, '81	25,26,31	Mononuclear leukocyte cultures
Conover et al, '73	30	Long-term lymphoblastoid cell lines
Wilson et al, '78	26	
Wilson et al, '80, '81	25, 27, 31	M–Mφ cultures
Wilson et al, '80, '81	25,27,31	T lymphocyte cultures.
Beratis et al, '73	32	Skin fibroblast cultures
Beratis et al, '73	32	Amniotic fluid cell cultures

*Also possibly in saliva. See Ref. 6.

linked to several of the biochemical, pathological, or clinical anomalies demonstrated in CF (Figure 2). Known properties of the CDS produced by purified M–Mφ and T lymphocytes from CF genotypes are summarized in Table 5. Most of these CDS may also be found in serum from CF genotypes, and at least three of them may be found in cell cultures or body fluids of CF genotypes.

The presence of the CDS in body fluids and cell cultures from CF genotypes may be due to an abnormal accumulation of normal products secondary to the primary metabolic block. [7,10,15,24] Nadler and co-workers[10] and Conover et al[33,34] assume that CDS accumulation results from a deficiency of an enzyme (either a proteolytic enzyme with trypsin-like activity or a carboxypeptidase) responsible for catabolism of these products. Our group[15,24,35] maintains, however, that an anomaly in the function or clearance of alpha$_2$-macroglobulin (A$_2$M), a broad spectrum enzyme regulator, is primary and leads to alterations in the level or functional state of key enzymes involved in CDS metabolism (Figure 2).

Evidence that both proteolytic enzyme inhibitors or regulators (A$_2$M) are involved in the cellular metabolism of the CDS (and CFP?) has been obtained recently using purified M–Mφ cultures. The results of these experiments also provide evidence that CDS accumulation in M–Mφ cultures from CF genotypes can be "corrected" by the addition of normal products made by M–Mφ.[24,36,38] Additional evidence demonstrates that under certain conditions normal M–Mφ cultures can be induced to accumulate CDS.

TABLE 5

Properties of CDS Produced by M–Mφ and T Lymphocytes from CF Homozygotes and Heterozygotes*

Property	CDS produced by M–Mφ			CDS produced by T lymphocytes
	C5 fragment	C3 fragment	Monocyte-CF specific CDS	T cell lymphokine-CF specific CDS
Approximate M.W.	15,000	9,000	5,000	5,000
Thermal lability (56°C, 45 min.)	stable	partially destroyed	destroyed	destroyed
Isoelectric point	5.9–6.6	8.1–8.6	2.2–4.6	6.9–7.8
Generation blocked by cycloheximide	yes	yes	yes	yes
Destroyed by pronase	yes	yes	yes	yes
Binds to IgG**	no	yes & to C3b	uncertain	yes
IgG needed for activity	no	no	no	no
Found in serum	yes	yes	uncertain	yes
Binds to concanavalin A+	yes	no	no	no
Neutralized by antibody C3	no	yes	no	no
Neutralized by antibody C5	yes	no	no	no
Present in cell cultures:				
normal controls++	no	no	no	no
patient controls	yes	no	no	no

TABLE 5, continued

Effects	Promote ciliary dyskinesia and mucus expulsion.	Promotes ciliary dyskinesia, modulates (inhibits response) of monocytes and PMN to chemoattractants.
	Chemoattractant for PMN and monocytes.	Possible chemoattractant for monocytes; less active for PMN.

*For further details see text and references 14, 24-27.
**Removed by protein A Sepharose—CL4B after incubation with human IgG.
+Failure to bind indicates either lack of carbohydrate or lack of accessible moieties.
++Based on biological activity and physicochemical properties such as charge, molecular weight, and association with IgG only. Immunochemical studies have not yet been completed for all of the CDS.

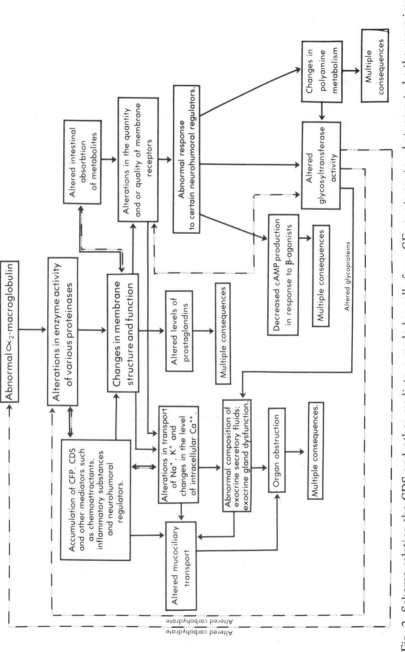

Fig. 2. Scheme relating the CDS and other mediators made by cells from CF genotypes to what seem to be the more important clinical, biochemical, or pathophysiological findings in CF. Dotted lines denote "feedback loops". Double headed arrows show interrelationships where the anomalies noted are mutually caused by or can modify one another. (Modified from Ref. 15.)

IMMUNOASSAYS FOR CFP

Development of Antisera and Qualitative Immunoassays for CFP

In order to develop qualitative or quantitative immunoassays for CFP, the first task was to obtain xenogeneic polyvalent antisera to CFP. BALB/c mice were studied for this purpose because success in raising mouse antisera to CFP would indicate the feasibility of producing large quantities of hybridoma antibody.[23] Other problems such as determining the proper immunization schedule, route of injection, and best physical state of the immunogen for the production of high quality antisera, could be determined cheaply using this species. The information obtained could be applied to the production of antisera using larger mammals.[23]

Our initial attempts to develop antisera to CFP entailed injecting mice with the CFP-containing region of the electrophoretogram obtained from focusing either whole serum, serum components desorbed from protein A, or serum components not bound to protein A[15] (Figure 3). In each instance the region in which CFP focuses was excised from the unstained electrofocused gel, emulsified in either phosphate buffered saline (PBS, pH 7.4) or sodium acetate buffer (pH 4.7), and injected intraperitoneally. To obtain hyper-immune antisera, mice were immunized up to seven times over a five month period.

Counterimmunoelectrophoresis (CIE) employing an agarose gel with medium electroendosomotic properties and barbital buffer (pH 8.6, ionic strength 0.10)[23] was first used to screen the various mouse antisera for re-activity. CIE and the conditions noted were chosen for several reasons. First, CIE requires the use of small quantities of antiserum. Second, electrophoretic runs can normally be completed in 90–180 minutes. Third, we had determined[39] by analyses of mouse sera employing IEF-TLPG followed by crossed immunoelectrophoresis, that mouse IgGs have a pI range of 5.0 to 7.7 (Figure 4). Thus, CIE at pH 8.6 is ideal for analyses of mouse antibody reactivity to CFP or other cationic human serum components. At pH 8.6 mouse antibodies will migrate toward the anode and will react only with the most cationic human serum components, such as CFP, which should migrate slowly toward the cathode.

Analyses of antisera from mice immunized with each of the three im-munogens indicated that only mice which had received the first two immu-nogens (Figure 3, A and B) produced antibodies recognizing additional com-ponents in CFP positive as compared to CFP negative samples (Figure 5, A–C). These results are consistent with our contention that CFP absorbs to protein A (Table 1). In addition, analyses of the mouse antisera produced indicated that emulsification of the excised gel in pH 7.4 PBS buffer resulted in more potent antisera than emulsification in pH 4.7 acetate buffer.[23] These results are con-sistent with a low molecular weight (M.W.) for CFP (<15,000)[14], and indicate that CFP should be a better immunogen when bound to a macromolecular

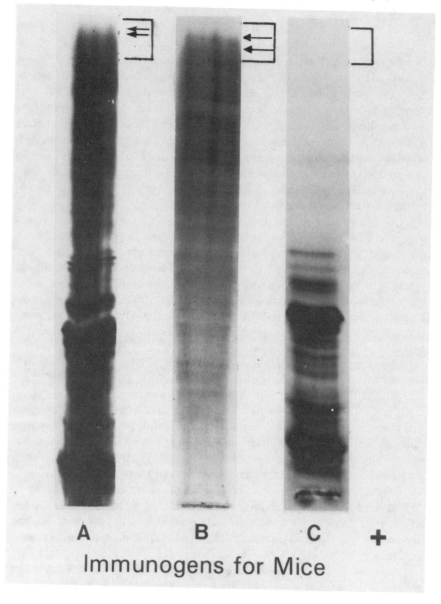

Fig. 3. Immunogens used to obtain anti-CFP antibodies. A = whole serum. B = serum components adsorbed to protein A. C = serum components not bound to protein A. The area of the electrophoretogram used is enclosed in brackets. Arrows indicate the position of the CFP doublet.

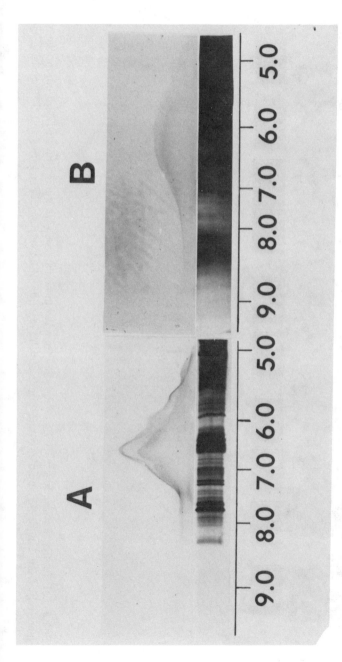

Fig. 4. Determination of the pI range for mouse IgG by IEF–TLPG followed by crossed immunoelectrophoresis. A = results when mouse serum was focused in a pH 2.5 to 10 gradient containing 4 M urea. B = results for a pH 2.5 to 10 gradient without urea. First dimension anode to the right (scale shows pH gradient). Second dimension anode at the top. A purified goat IgG fraction containing anti-mouse IgG antibodies was used to obtain the pattern shown.

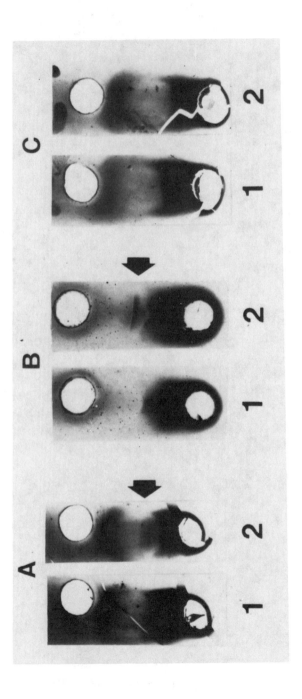

Fig. 5. CIE results showing reactivity of unabsorbed immune mouse antisera from mice injected with immunogens A, B, and C respectively from Figure 3. Numbers 1 and 2 indicate the reaction obtained for CFP negative and CFP positive human sera.

TABLE 6

Correlation Between Serum CFP Positivity Determined by IEF and Serum
Reactivity Determined by CIE Using Mouse Antisera Absorbed with CFP
Negative Normal Human Serum

Serum	No. anal.	No. + CFP by IEF	No. + CFP by CIE*
CF homozygote	11	11	11
CF heterozygote	12	10	11
Normal control	14	0	1

*The increased frequency of reactive samples shown by CIE as compared to IEF
may be due to the fact that CIE is a more sensitive method than IEF.

carrier such as IgG. (From the results summarized in Table 1, it would appear
that CFP is dissociated from its carrier at pH 4.7). We are currently attempting
to obtain more potent antisera to CFP by coupling[40] partially purified preparations of CFP to macromolecular carriers before injection. (All of the results
to be reported in this communication are from experiments using uncoupled
immunogens.)

In our analyses of human sera by CIE using unabsorbed mouse antisera
to immunogen A or B (Figure 3), we found that CFP positive samples produced more precipitation arcs than did CFP negative samples (Figure 6). To
produce monospecific antisera to CFP, we initially resorted to removing un-
wanted antibodies to components common to both CF and normal sera by
liquid-phase absorption of immune mouse sera with a pool of CFP negative
normal human sera. This method was successfully employed in three out of six
attempts to create antisera which recognized components in CF but not normal
sera. The results obtained from our sera analyses of 11 CF homozygotes, 12
heterozygote carriers, and 14 normal controls, using one of the absorbed antisera, are summarized in Table 6. These preparations of absorbed antisera produced one major and at times one obvious minor arc against CFP positive sera.
Normal control (CFP negative) sera usually failed to show any precipitation
arcs (Figure 7). Quantitative differences were also evident between the levels
of the components in CF homozygote and heterozygote carrier sera as evidenced by the location of the arcs relative to the cathode.[23]

The absorbed antisera obtained by the above method were considerably
less potent than unabsorbed antisera. For three preparations, absorption with
normal control sera had apparently resulted in the complete elimination of
reactivity for CF or normal samples. (No reaction was observed when a volume
20-fold greater than that used for unabsorbed antisera was employed. The three
reactive preparations required five to ten-fold more antisera than analogous unabsorbed preparations to demonstrate precipitation arcs.) These results sug-

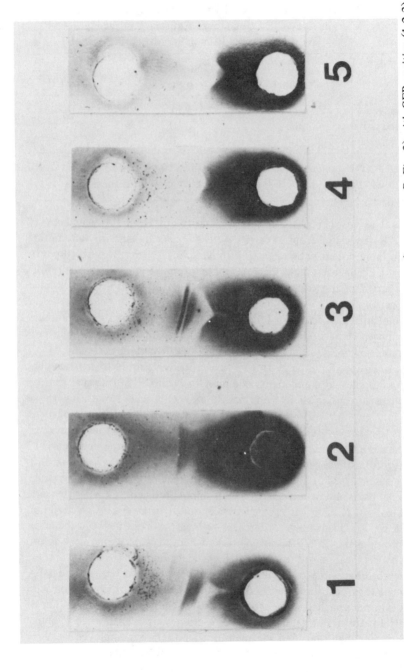

Fig. 6. CIE results showing reactivity of unabsorbed immune mouse sera (immunogen B, Fig. 3) with CFP positive (1,2,3) and CFP negative (4,5) sera. Note presence of additional components in CFP positive as compared to CFP negative samples.

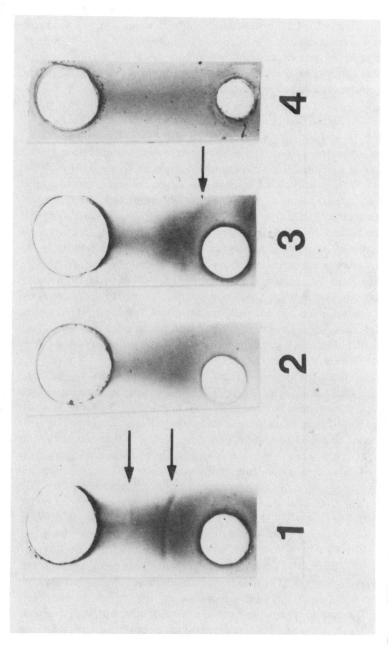

Fig. 7. CIE results showing reactivity of immune mouse sera previously absorbed with a pool of CFP negative control sera.
1 = CF homozygote sera. 3 = CF heterozygote carrier sera. 2 & 4 = normal control CFP negative sera.

gested to us that normal sera may either innately contain components or generate components during the absorption process which show at least partial antigenic cross reactivity with CFP. Before attempting to develop more quantitative immunoassays for CFP we felt we should determine which of the above alternatives was more likely and learn more about the identity of CFP. Accordingly, we sought to determine which serum components reacted with unabsorbed mouse antisera, and which serum components could be identified in extracts of electrofocused gel segments containing CFP.

Identification of Serum Components which React with Mouse Antisera

Figure 8 (A–D) shows typical results obtained from our analyses of the reactivity of unabsorbed mouse antisera to immunogen A or B (Figure 3) with either CF or normal sera using immunoelectrophoresis (IEP) at pH 8.6 (barbital buffer). With both types of antisera, one major arc was obtained against either CF or normal serum. This component was identified as IgG by subsequent analyses using Ouchterlony double immunodiffusion with both our mouse antisera and commercial antisera. Of greater interest, however, was finding additional antibody reactivity to other anodally and cathodally migrating serum components as evidenced by the presence of bifurcations or flares coming from the major IgG arc (Figure 8). A total of two to four anodal arcs were present depending on which types of mouse and human sera were studied. In general, mice injected with immunogen A (Figure 8, B) recognized more anodal components than mice injected with immunogen B (Figure 8, A). In addition, CF sera usually generated at least one more anodal arc than normal sera (Figure 8, B), or the arcs or bifurcations present deviated to a greater extent from the major IgG arc (Figure 8, D). Additional cathodal arcs (Figure 8, A,C) were noted when CF serum was analyzed, and were much more obvious when mice were immunized with immunogen B (Figure 3). From our results using IEP, it would appear that at least five different components may be present in the section of gel containing CFP and that one of them is IgG. Alternatively, fewer components may actually be present but may cross react with other serum components which share common antigenic determinants.

The finding of an additional cathodal arc in CF serum might be expected from our IEF-TLPG results. The fact that this secondary cathodal arc fuses with the major IgG arc in CF serum also seems consistent with our hypothesis that CFP is a low M.W. serum component which complexes to IgG (Table 1). Apparently, antibodies are present which recognize both IgG and the CFP-IgG complex. The presence of the additional anodal arcs seems to indicate two possibilities: (a) CFP and other cationic components in extracts of the pI 8.5 region might be fragments of larger globulins with beta or alpha$_2$ electrophoretic mobility which are released during electrofocusing in 4 M urea. (b) Mouse antibodies raised against these fragments must be capable of reacting with these peptides when they are still part of their parent molecules.

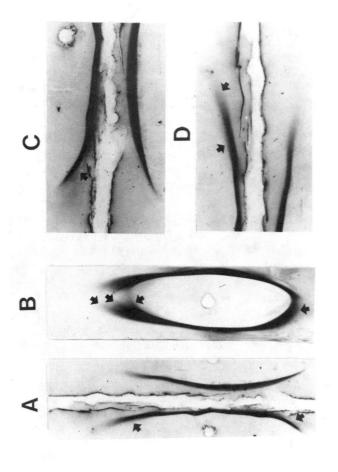

Fig. 8. Analysis of reactivity of mouse antisera and CF or normal human serum by IEP. A = results for mouse antisera generated by injecting mice with immunogen A (Fig. 3). B = results for mouse antisera generated by injecting mice with immunogen A (Fig. 3). C and D = enlargements of the bottom and top halves of the pattern shown in A. Anode at top in A and B, to the right in C and in D. Arrows denote the presence of bifurcations of spurs (arcs) emanating from the major arc, due to a reaction between human IgG and the mouse antisera. CF serum is on the left and normal serum is on the right in A and B.

Our results from analyses of extracts of gel segments containing CFP provided support for both possibilities.

Concentrated extracts of gel segments containing CFP (Figure 3, A) demonstrated at least four precipitation arcs when tested against unabsorbed mouse antisera (immunogen A) using CIE (Figure 9, A). Fractionation of the extract by Bio-gel P–10 chromatography,[26] followed by analyses of concentrated column fractions by CIE using the same mouse antiserum, indicated that three of these four cationic components were of M.W. 3,500–15,000 (Figure 9, B). The other component(s) eluted in the column void volume (M.W.>17,000) which was found to contain IgG by Ouchterlony double immunodiffusion using commercial antisera to normal human serum and IgG. Using Ouchterlony double immunodiffusion and monospecific goat or rabbit antisera to various human serum components, we have also succeeded in identifying fragments of two other serum components in the gel extracts (Figure 10): human complement component C3 and A_2M. Intact C3 is a globulin with beta electrophoretic mobility whereas A_2M is a macroglobulin with $alpha_2$ mobility. In separate analyses, we have confirmed that two of the anodal flares observed by IEP do indeed result from mouse antibodies which react with intact C3 and A_2M.

We continue our attempts to determine whether additional serum components are represented by fragments in the gel extract and, if so, to identify them. At this time we cannot determine whether or not CFP is a fragment of C3 or A_2M. However, our results to date support the contention that CFP is similar to the C3 fragment in the extract, in that both of these substances seem to be cationic fragments of beta or $alpha_2$ globulins which can bind to IgG. The CFP "doublet" (Figure 1) possibly includes both these components. Also, most likely the additonal cathodal arc observed for CF serum by IEP does result from mouse antibodies recognizing either the CFP or C3 fragment IgG complex, whereas one (or two) of the anodal flares from the major IgG arc does result from antibodies which recognize the "parent" of CFP or the C3 fragment and IgG. Further evidence supporting the possibility that CFP could be a C3 fragment, or that at the very least two potential markers for CF genotypes focus near pI 8.5, is presented in a subsequent section of this communication.

Both explanations proposed above for the extreme loss of potency of our mouse antisera after absorption with normal serum may be operable. Normal serum does apparently harbor a beta or $alpha_2$ globulin antigenically similar to the parent molecule of CFP present in CF serum, and fragments antigenically similar to CFP may be generated from proteolysis of the parent molecule during absorption. However, normal serum processed as described[12-14] does not contain significant amounts of components similar to either, "free" or IgG complexed CFP, both of which would be expected to migrate toward the cathode during CIE and react with the mouse antisera. To explain the fact that three preparations of monospecific antisera could be generated after absorp-

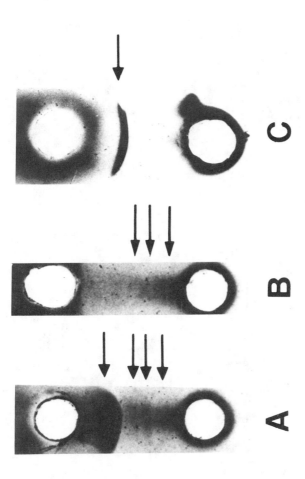

Fig. 9. Results of analysis by CIE of reactivity of mouse antisera prepared by injecting mice with immunogen A (Fig. 3). A = components extracted from the same region of the electrophoretogram used to prepare immunogen A. C = components of M.W. >17,000 isolated by Bio-gel P–10 chromatography[25] of the extracted components in A. B = components of M.W. 3,500–15,000 isolated by Bio-gel P–10 chromatography of the extracted components in A. Arrows indicate the presence of components precipitated by the mouse antisera.

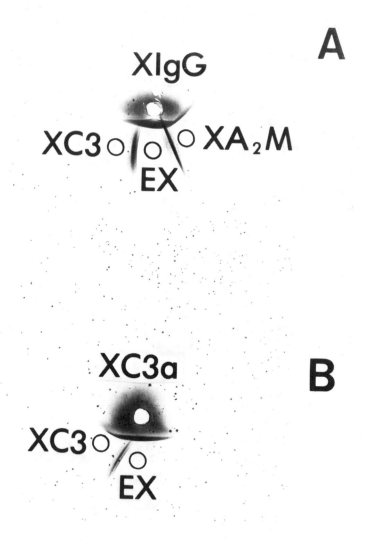

Fig. 10. Demonstration of the presence of components which may be fragments of human complement component C3 and A_2M in extracts of gels from the region of electrofused serum known to contain CFP (Fig. 3,A) by Ouchterlony double immunodiffusion. Ex: concentrate of the extracted components. XIgG, XC3, XC3a, XA_2M: goat or rabbit antisera to human IgG, C3, C3a and A_2M respectively.

tion with normal human serum, we propose that those antibodies that recognize CFP-IgG complexes are not completely removed by crossreacting antigenic determinants present in the intact normal parent molecule or its normal fragments, especially if the unabsorbed anti-CFP mouse sera initially strongly recognized antigenic determinants unique to the CFP-IgG complex.

Development of a Quantitative Immunoassay for CFP

To quantify more accurately the differences in concentration of CFP in sera of CF homozygote and heterozygote carriers suggested by our results using CIE[23] (Figure 7), we developed a rocket immunoelectrophoresis (RIEP) assay. The technical problems inherent to RIEP were initially solved using mouse antibody to human IgG and normal human serum. Since we had previously determined that mouse IgGs have a pI range of 5.0–7.7, we knew that RIEP had to be performed at a slightly acidic pH to minimize the mobility of mouse IgG. Preliminary experiments indicated that a pH of 6.5 was optimal. At this pH we could readily quantitate the level of human IgG in serum (Figure 11, A), and we obtained at least three rockets for CFP positive samples (Figure 11, B). However, unabsorbed mouse antisera failed to produce rockets specific for CF genotypes when whole sera were studied. Based on the information obtained from our attempts to identify serum components recognized in gel extracts by our mouse antisera, and clues provided by conducting other related experiments, we determined that the major reasons for the lack of specific rockets for CF genotypes using unabsorbed antisera were:

1. Specific components (*e.g.*, CFP and CFP-IgG complexes) were being obscured by the presence of cathodal rockets apparently resulting from both unrelated components and parent molecules migrating toward the cathode at pH 6.5. This was not a problem when CIE was employed, since at pH 8.6 serum components with beta or alpha$_2$ electrophoretic mobility migrate toward the anode.
2. High concentrations of antisera are required. This was not too surprising; in CIE, an antiserum concentration of 100% (undiluted) was employed.
3. Point one was compounded by a problem realized earlier that normal samples most likely contain a serum component which is antigenically crossreactive with the CFP parent with beta or alpha$_2$ electrophoretic mobility.

To counteract these problems, we resorted to:

1. The use of antisera rendered selectively polyspecific by removing antibodies to specific serum components using solid phase absorption or monospecific by absorption with specially treated normal control sera (CFP negative by IEF-TLPG).

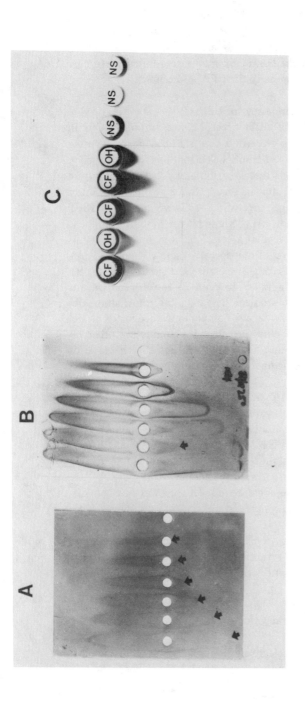

Fig. 11. Analysis of reactivity of mouse antisera by RIEP. A = plate showing quantitation of human IgG using 1.0% mouse anti-human IgG antiserum. Arrows show tip of cathodal rockets. Twenty μl of normal serum was cut 1/10 and serially diluted to 1/320. B = unabsorbed mouse antisera (2.5%) (immunogen B, Fig. 3). Wells contained 20 μl human serum (CFP-positive) diluted 1/5 (first two wells) then serially diluted to 1/80. Arrow denotes a second cathodal rocket not seen in A. Far right well in A and B is a buffer-only control. C = quantitation of material believed to contain CFP, obtained from sera of CF homozygotes (CF), CF heterozygotes (OH), and normal controls (NS). The gel contained 25% absorbed anti-CFP mouse antisera. Buffer was sodium barbital-sodium acetate, pH 6.5 (ionic strength 0.02). Cathode at the bottom.

2. The development of methods to obtain CFP (and other potentially relevant markers) free of contamination from other serum components.
3. The inclusion of substances to stabilize serum samples.

We now routinely employ all of the above in analyses using either CIE or RIEP.

After investigating the above problems, quantitative differences could be seen between CF homozygote, heterozygote carrier, and normal control samples by RIEP (Figure 11, C). More recently, we have succeeded in increasing the sensitivity of our immunoassays by using ^{125}I-labeled reagents.[39] Experiments using mouse antisera of a potency similar to that employed in regular RIEP studies (Figure 11, C) indicate that we can use as little as 0.5% monospecific antiserum with radio-RIEP (Figure 12).

Results obtained using both regular and radio-RIEP indicate that CFP levels in samples from CF homozygote, heterozygote carrier, and normal controls are sufficiently different to distinguish the three genotypes (Table 7). Studies are still in progress to determine the specificity of CFP for the CF gene as determined by RIEP. However, our results so far, using sera from one patient control with chronic allergy and two patient controls with chronic pulmonary disease, (Table 7) indicate that CFP levels are sufficiently higher in heterozygote carriers to utilize RIEP as a diagnostic assay for heterozygote detection. In addition, attempts are currently being made to use radio-CIE and radio-RIEP to detect and quantitate CFP in white blood cell and skin fibroblast cultures, and in amniotic fluid, as a prelude to setting up a CFP assay for prenatal diagnosis. Our preliminary findings indicate that CFP is present in culture supernatants from mononuclear leukocytes. This result is consistent with the possibility (based on the finding of CFP in sera from leukemic patients) that CFP accumulates in sera due to a metabolic defect involving leukocytes.

TABLE 7
CFP Levels in Sera as Shown by RIEP

Sample	No.	Rocket height (mm)*	
		Values	Mean ± SD
CF homozygotes	8	13, 11, 15, 21, 9, 18, 17, 23	15.90 ± 4.80
CF heterozygotes	6	5, 3, 7, 6, 2, 8	5.17 ± 2.32
Normal controls	8	0, 0, 0, 0, 0, 1, 0, 0.5	0.19 ± 0.37
Patient controls**	3	0, 0, 0.5	0.17 ± 0.29

*Values are corrected for varations between runs by including at least one common sample on all electrophoresis plates.
**Patient controls included one non-CF patient with chronic allergy and two with pulmonary disease.

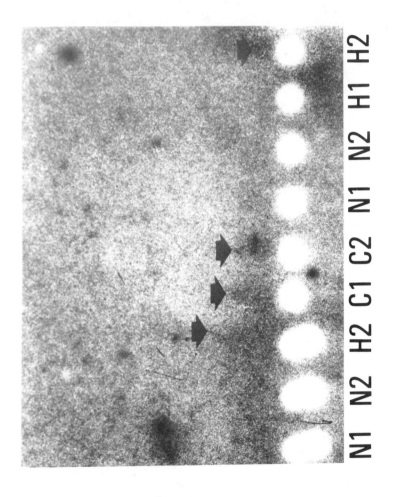

N1 N2 H2 C1 C2 N1 N2 H1 H2

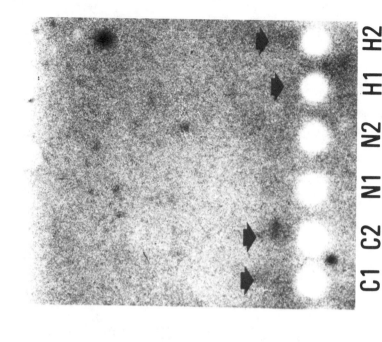

C1 C2 N1 N2 H1 H2

Fig. 12. Top. Quantitation of material containing CFP from sera of normal controls (N), CF heterozygotes (H), and CF homozygotes (C) by radio-RIEP. The gel contained 0.5% mouse anti-CFP antiserum. The single wells contained one-third as much material as the double wells. Note that both normal control samples failed to react with the mouse antiserum. The tips of the cathodal rockets in reactive samples are indicated by arrows.

Bottom. Enlargement of wells 4–9.

TABLE 8

Neutralization of CDS by Nonimmune and Immune Mouse Sera

CDS tested*	Neutralized by (lowest concentration produces effect)	
	Immune serum**	Nonimmune serum
C5 fragment	No (1/10)	No (1/10)
C3 fragment	Yes (1/50)	No (1/10)
M-Mϕ CF-specific CDS	Possible reduction in activity (1/10). No effect when diluted beyond 1/10.	No (1/10)
T lymphocyte CF- specific CDS	Possible reduction in activity (1/10). No effect when diluted beyond 1/10.	Same as immune serum.

*See Table 5 for further description of each activity. Effects noted are for CDS purified by IEF-TLCG[13] of concentrated culture supernatants.
**Mice immunized with immunogen B (Fig. 3).

STRUCTURAL RELATIONSHIP BETWEEN THE CILIARY DYSKINESIA SUBSTANCES AND CYSTIC FIBROSIS PROTEIN SHOWN BY IMMUNOLOGICAL TECHNIQUES

The possibility that CFP is structurally similar to one of the CDS (Table 5), particularly the C3 fragment CDS, seems evident from the results we have reviewed concerning the possible identity of CFP, and from striking similarities in physicochemical properties noted for CFP (Table 1) and the C3-CDS (Table 5). To link CFP more firmly to the C3-CDS or another CDS, we attempted to neutralize the various CDS activities purified by Bio-gel P-10 chromatography[25] and isoelectric focusing[27] by incubating them with mouse antisera to immunogen B (Figure 3). The mouse antisera used in these experiments were first heat inactivated (56°C, 45 min.), rendered nonreactive to human IgG by solid phase adsorption employing human IgG covalently linked to A–H Sepharose 4B,[41] and then exhaustively dialyzed against PBS using dialysis tubing with a M.W. retention limit of 20,000. Mouse sera obtained from nonimmune BALB/c mice were treated in a similar fashion and used as a control (Table 8).

Only the C3 fragment CDS was consistently neutralized by immune mouse sera diluted beyond 1/10. The effects observed for the T lymphocyte derived CDS were nonspecific since nonimmune mouse sera produced similar effects.

The reduction in activity observed when the M-Mϕ derived CF-specific CDS was incubated with immune serum may reflect partial antigenic identity of the CDS component in this fraction (pI 2.2–4.6) with one which focuses in the pI range of CFP. Alternatively, this may indicate that the M-Mϕ derived CDS activity results from the combined effects of more than one component (possibly subfragments of the C3 fragment and C5 fragment CDS), and that only one of them is neutralized by the mouse antisera. The results in Table 8 are entirely consistent with the known pIs of the CDS and add further support to the possibility that CFP may be the C3 fragment CDS. It remains possible, however, that CFP is not the C3 fragment CDS; both components may simply exist in the gel segment which contains CFP. Further experiments are currently being conducted to resolve this. Our results do indicate, however, that we now have antibodies to two markers for CF genotypes.

Regarding the specificity of the C3-CDS, it has been shown by Conover et al[42] and confirmed by us[43] that sera from non-CF patient controls with agammaglobulinemia, bronchial asthma, coeliac disease and chronic allergy harbor a CDS. Until recently[25] both Conover et al[34] and we[14,26] assumed that the CDS found in sera from non-CF patient controls was C3a anaphylatoxin, a 9,000 M.W. fragment of C3.[44] This assumption was based mainly on evidence that samples from CF patients contained a CDS which was neutralized by antisera specific for human C3a,[31,34] or that fractions obtained from gel filtration chromatography of CF serum containing a CDS of M.W. near 9,000 reacted immunologically with antisera to human C3a.[14] Additional circumstantial evidence that the CDS in CF serum was C3a was provided by Conover et al[34] in that the CDS activity in CF serum could be inactivated by carboxypeptidase B and that a similar activity could be generated in normal sera after treatment with epsilon-amino-caproic-acid (EACA). However, carboxypeptidase B inactivates C3a as well as C5a, and C5a is also generated by EACA treatment.[44] In addition, C5a has a M.W. (11,000) quite close to that of C3a.[44] The crucial experiment to demonstrate the existence of C3a with CDS activity in samples from patient controls was never performed by our group or Conover's.

Recently, during the course of an investigation designed to elucidate whether or not the CDS might be chemoattractants for polymorphonuclear (PMN), we did attempt to determine if MNC from patient controls with pulmonary disease synthesize a CDS which reacts with antiserum to human C3 or C3a.[25] While we could demonstrate that MNC's from CF genotypes and patient controls do generate a CDS of M.W. near 9,000 (as determined by Bio-gel P-10 chromatography) which is chemotactic for PMN,[25] the chemoattractant activity and the CDS activity in patient control samples was found to be a fragment of C5 not C3. Normal MNC cultures also generated a C5 fragment PMN chemoattractant which did not have ciliary dyskinesia activity, while fractions from CF genotypes harbored fragments of C3 and C5 which were CDS as well as PMN chemoattractants.[25] Recently, we have extended these results by demonstrating that MNC culture supernates from patient controls do not contain a CDS or PMN chemoattractant activity in the pI 8.1–8.6 fraction obtained by IEF-TLPG which is believed to contain the C3 fragment CDS (Table 5). We conclude that patient control samples contain a C5 fragment CDS but not a C3 fragment CDS. Moreover, we conclude that the C3 fragment CDS found in samples from CF genotypes is not active C3a anaphylatoxin; C3a is not chemotactic for PMS[44] whereas the C3 fragment CDS is chemotactic for PMN[25] and also MNC's.[27] Their pI's (8.1–8.6 vs 9.7:44) are also different.

METABOLIC CORRECTION OF CDS ACCUMULATION *IN VITRO*

As indicated earlier, two hypotheses have been proposed to explain the accumulation of the CDS or CF factors in cell cultures and body fluids from CF genotypes. One proposed by Conover *et al*[34] and Nadler and coworkers[10] suggests that CF genotypes have an innate defect in an enzyme which functions to inactivate the CDS. Nadler and coworkers feel that this enzyme is a serine esterase or serine proteinase with trypsin-like activity.[10] Their proposal is based on their findings that CF saliva and serum are deficient in a trypsin-like enzyme which can be readily demonstrated in analogous normal samples.[10] However, it has never been demonstrated whether or not the enzyme found in normal control samples can actually inactivate any CF factor. Conover *et al*[34] suggested that a carboxypeptidase B-like enzyme (anaphylatoxin inactivator[44]) was deficient in CF, on the basis of presumptive evidence that all of the CDS activity in CF serum could be attributed to C3a-anaphylatoxin and could be inactivated by carboxypeptidase B. However, no deficiency in anaphylatoxin inactivator activity could be demonstrated in CF serum by other investigators (reviewed in Ref. 10). Our own results from MNC derived CDS activities[25] indicate that there are at least three CDS, none of which is C3a *per se*.

We have proposed another theory[35] which maintains that A_2M, a glycoprotein produced by M-Mϕ[36,37] found in plasma and other body fluids, and is a broad spectrum proteinase inhibitor, is structurally altered in CF – or that its active site for enzyme binding is modified or blocked endogenously by polycations, which may be generated in higher amounts in CF than in normal body fluids. The major tenet of our hypothesis is that an abnormality related directly to the function or clearance of A_2M leads to an alteration in the level or functional state of key enzymes involved in the metabolism of the CDS. In other words, we feel that the enzyme defect proposed by Nadler[10] is secondary to a metabolic anomaly involving A_2M.[25] Evidence that the carbohydrate portion of CF A_2M is structurally abnormal,[45] and that polycations can alter its proteinase binding activity[46] has been published. Whether or not there is an innate defect in the enzyme binding site of CF A_2M is controversial.

It is known that clearance of A_2M from the body fluids is performed predominantly via endocytosis by cells of the reticuloendothial system (M-Mϕ or their derivatives) only when A_2M is bound to an enzyme. Free A_2M is not normally endocytosed.[47] When A_2M binds a proteinase the esterolytic but not the proteolytic activity of the enzyme is maintained.[48] Thus, if A_2M-enzyme complexes persist in the circulation, they may modify both cellular and circulating body components.[48] In addition, recent evidence suggests that A_2M-enzyme complexes (but not free A_2M or complexes of other proteinase inhibitors with enzyme) can suppress M-Mϕ cellular activity,[47,49] and when A_2M-enzyme complexes are endocytosed by M-Mϕ there is a selective release of enzymes with neutral proteinase activity.[50] This latter finding is particularly important since it offers an explanation for an enzyme deficiency in cell cultures or body fluids from CF genotypes. If endocytosis of CF A_2M is diminished due to some anomaly associated with CF A_2M, then reduced enzyme release might occur resulting in an enzyme deficiency.

Over the past three years, we have been studying CDS metabolism using isolated M-Mϕ as a model.[51] M-Mϕ were selected since CF M-Mϕ generate at least three CDS (Table 5) and also potentially secrete components such as C3, C5, proteinases, alpha$_1$-antitrypsin (A_1AT), and A_2M which may be involved in CDS metabolism and are now known to be released by normal M-Mϕ.[36-38,52-54] It is known, for example, that two of the CDS are fragments of C3 and C5 and that fragments of C3 and C5 with chemotactic activity for PMN and M-Mϕ can be generated by proteolytic cleavage of C3 and C5 by M-Mϕ secreted enzymes.[53,55] The enzymes involved are normally inhibited or regulated by A_1AT or A_2M.[48,56] It seemed possible therefore that we might be able to deduce the metabolic abnormalities responsible for CDS accumulation by comparing CF and normal M-Mϕ metabolism *in vitro*.

Direct evidence that both proteinases and proteinase inhibitors are involved in the cellular metabolism of the CDS has now been obtained by our group[51] (G.B. Wilson and E. Floyd, in preparation). In one phase of these studies (Table 9) proteinases and proteinase inhibitors were added to cultures of purified M-Mϕ to determine their effects on the generation of the CDS. The absence of CDS activity in cultures of CF M-Mϕ supplemented with the serine proteinase inhibitors A_1AT or soybean trypsin inhibitor (SBTI), and the increase seen with the serine proteinases trypsin and plasmin indicate that the CDS are generated when macromolecules, such as C3 and C5 normally made by M-Mϕ,[54] undergo proteolytic cleavage. Inadequate regulation of enzyme activity either by A_1AT or more likely by A_2M, since A_1AT may transfer its bound enzyme(s) to A_2M,[47] could result in excessive proteolytic activity. A more complex mechanism is suggested by our findings: (a) Normal M-Mϕ make CDS when trypsin, plasmin, or normal A_2M is added. (b) Normal A_2M complexed to trypsin or plasmin (exogenously before addition to the cultures) inhibits CDS generation by CF M-Mϕ. (c) Normal M-Mϕ generate complement fragments (particularly the C5 fragment) without dys-

TABLE 9
Summary of Effects of Enzyme Inhibitors, Enzyme Regulators,
and Enzymes on the Generation of CDS in Cultures of Purified M-Mφ*

Substance**	Conc. range (ng/ml)	CF genotypes	Nl controls
None			
Culture medium			
(control)	NA	Yes[+]	No
Proteinase inhibitors			
A_1AT	0.25–250	No	No
SBTI	0.10–100	No	No
Normal A_2M	5.00–500	Yes, increased	Yes
CF A_2M	5.00–500	Yes	Yes (at hi conc)[++]
Proteinases			
Trypsin	0.1 –100	Yes, increased	Yes
Plasmin	1.0 –500	Yes, increased	Yes
Complexed inhibitors [†]			
A_1AT or SBTI	Same as above	Yes	No
Normal A_2M	Same as above	No	No
CF A_2M	Same as above	Yes	No

*M-Mφ were purified and cultured in RPMI–1640 medium containing 1% bovine serum albumin and antibiotics as described in detail elsewhere.[25,31] Each culture contained $2.0–5.0\times10^5$ M-Mφ at the beginning of the experiment. Cultures were maintained for 96 hours. All the effects noted were concentration dependent.

**Substances were added at 0, 24, 48 and 96 hours at the final concentrations (in culture) indicated. Each substance was purified to "functional homogeneity" (no other inhibitors or enzymes present) and tested for activity before use. None of the substances had direct effects on ciliary movement in the rabbit tracheal bioassay used to detect CDS activity.[31].

+Yes = CDS generated. No = No CDS activity generated.

++At least 100 to 500 ng/ml CF A_2M were required to generate some CDS activity. Effects with normal A_2M were observed with as little as 5 ng/ml.

†Complexed to trypsin or plasmin of a molar ratio of one to one using a concentration range equal to that employed when the proteinases or proteinase inhibitors were tested alone.

kinesia activity.[25] The first finding seems to preclude an enzyme deficiency involving "trypsin-like" serine proteinases, since clearly the addition of serine proteinases to normal M-Mϕ cultures leads to CDS accumulation.

We believe that these findings indicate that CDS accumulation is a multi-step process. In the first step, a precursor such as C3 or C5 is cleaved by proteinase(s) (probably a serine or neutral proteinase, but possibly an acid proteinase such as cathepsin D[52,53]) to generate pre-CDS fragments. These fragments are then structurally modified in a second step which endows them with CDS activity. Catabolism of the products of the first step prior to modification, or of the CDS themselves, then occurs. This third step probably involves enzymes different from those responsible for the generation of CDS, because A_1AT and A_2M promote inverse effects (Table 9). A_2M can bind and inhibit all enzymes bound by A_1AT, but the inverse is not true.[48]

To explain the fact that addition of normal A_2M leads to CDS accumulation we propose two possibilities. First, A_2M may preferentially bind enzymes involved in CDS catabolism, which cannot be inhibited by A_1AT (e.g. acid proteinases). Second, enzyme-A_2M complexes formed endogenously may be active in creating the CDS, by modifying the pre-CDS produced in the first step. Here we would assume that the enzymes actively involved in this process are not serine proteinases per se, since normal A_2M exogenously complexed to serine proteinases leads to inhibition of CDS generation. These two concepts can be united by assuming that normal A_2M preferentially binds enzymes involved in pre-CDS or CDS catabolism, which when bound to the inhibitor can still modify the pre-CDS to create CDS but cannot inactivate either moiety. This is roughly analogous to the preservation of the esterolytic activity of serine proteinases when they are bound by A_2M.[48]

The disruption of CDS catabolism in normal M-Mϕ by normal A_2M probably occurs prior to endocytosis of A_2M-enzyme complexes, since the ultimate effect of A_2M-enzyme complexes (but not other inhibitor-enzyme complexes) on M-Mϕ is suppression of cellular activity.[47,49] Indeed, our results indicate that normal A_2M complexed to serine proteinases suppresses CDS generation by CF M-Mϕ (Table 9). The failure of CF A_2M preincubated with serine proteinases to inhibit CDS accumulation suggests that CF A_2M-enzyme complexes are not efficiently engulfed by CF M-Mϕ, or if engulfed are not suppressive. The failure to engulf CF A_2M-enzyme complexes to suppress CF M-Mϕ CDS accumulation does not appear to be due to abnormal receptors for A_2M-enzyme complexes on CF M-Mϕ since normal A_2M-enzyme complexes suppress CDS accumulation in CF M-Mϕ cultures. The problem seems to reside with CF A_2M and could be due either to (a) the abnormal carbohydrate structure of CF A_2M, or (b) the inability of proteases bound by CF A_2M to induce the structural modifications in CF A_2M necessary for endocytosis or suppression. Either mechanism may result in a failure of CF A_2M to suppress the metabolic activity of M-Mϕ.

To explore further the mechanisms behind CDS accumulation in M-Mϕ, we measured total protein, A_1AT, A_2M, C3, and C5 synthesis in M-Mϕ cultures from CF genotypes and normal controls (Figures 13 and 14). The findings are as follows:

1. CF homozygote M-Mϕ synthesize more protein than normal M-Mϕ at all time points evaluated, whereas CF heterozygote M-Mϕ synthesize more protein at 24 and 48 hours, and less protein at 72 and 96 hours than normal M-Mϕ. Although the amounts of protein synthesized by CF genotype M-Mϕ and normal M-Mϕ cultures are not consistently significantly different, the trend which is evident is also reflected in the results for C3, C5, and A_1AT synthesis but not A_2M (Figure 14). The difference in protein synthesis noted for normal control and CF genotype M-Mϕ seems to reflect changes in secreted protein moieties (Figure 14).

2. The fact that CF M-Mϕ secrete more of all four components studied than normal M-Mϕ suggests that CF M-Mϕ are both metabolically hyperactive and hypersecretory.

3. From the results for C3 and C5 synthesis, it appears that CDS accumulation is due to both hypersecretion of CDS precursors and hypercatabolism of these "precursors" (C3, C5) to produce the CDS.

4. It is possible that the A_2M made by CF M-Mϕ is abnormal (Figure 14). First, CF M-Mϕ produce significantly higher levels of A_2M than normal M-Mϕ yet CF cultures still accumulate CDS. Second, unlike the trends observed for CF M-Mϕ with respect to C5, C3, and especially A_1AT synthesis (which indicate that less product is synthesized at 48 hours than 24 hours, and less at 96 hours than 72 hours except for C3), A_2M synthesis actually shows a gradual increase from 24 hours to 96 hours (possibly an attempt by CF M-Mϕ to regulate their activity). The general trend, which is evident when CF heterozygote and normal control M-Mϕ cultures are compared with respect to C5, C3, A_1AT, and total protein synthesis is also strikingly different for A_2M. CF heterozygote M-Mϕ actually synthesize more A_2M than normal M-Mϕ at 72 hours. (Figure 14).

The above findings, and those in Table 9, add support to our contention that the primary cause of CDS accumulation by CF M-Mϕ may be an anomaly involving A_2M. Our results also indicate an innate defect in the ability of CF M-Mϕ to regulate their own activity; they appear both hyperactive and hypersecretory. This regulatory defect may be due to an abnormality in CF A_2M, however, other aberrations of regulation of M-Mϕ secretion or metabolic activity may also be involved. Included here are the possibilities that CF M-Mϕ also express defects in stimulus-secretion coupling mechanisms resulting from anomalies in intracellular levels of calcium[57,58] or abnormal responsive-

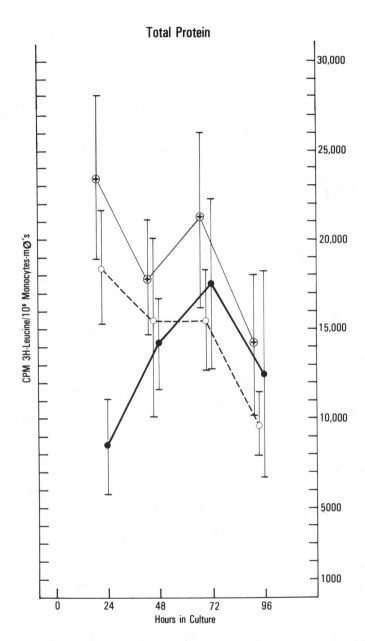

Fig. 13. Total protein synthesized by M-Mϕ from CF homozygote (+), CF heterozygote carrier (0- - -0), and normal control subjects (●——●). For details see Refs. 25, 38.

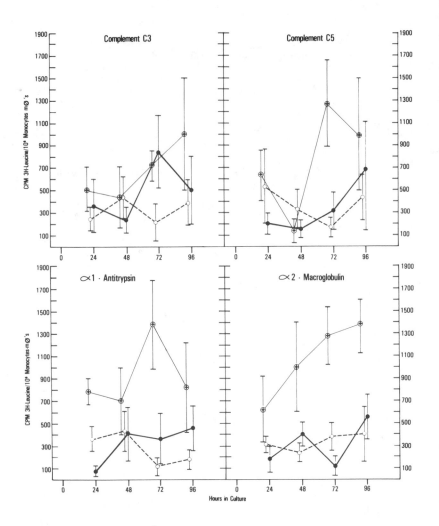

Fig. 14. Secretion of ^{3}H-labeled complement component C3, complement component C5, A_1AT and A_2M by M-Mϕ from CF homozygotes (+), CF heterozygote carriers (0- - -0) and normal controls (●——●). For details see Refs. 25, 38.

Author's note: Additional reference for Figures 13 and 14: Wilson GB. Ciliary dyskinesia factors produced by leukocytes. *Lymphokines* 1982: 8. In press.

ness to neurohumoral regulators (already shown in PMN[59]). Such defects have been proposed to explain exocrine gland dysfunction in general in CF,[57-59] and if operable *in vitro* or especially *in vivo* could be working in concert with a defect in A_2M to produce M-Mϕ which are innately metabolically hyperactive, and incapable of modulating their activity or the metabolism of mediators (the CDS and others) which can induce secretion. At least two of the CDS (the C3-fragment CDS and C5-fragment CDS) can induce M-Mϕ secretion[53,55,60] resulting in an escalation of their own production or accumulation. Once produced, the CDS can also induce alterations in intracellular levels of Ca++ and alterations in the activity of glycosyltransferases,[61] leading to changes in the carbohydrate structure of macromolecules such as A_2M, and may affect the cellular response to neurohumoral regulators as well. The multiple changes in cellular metabolism produced by the CDS (or other mediators) make it difficult to sort out cause and effect relationships (Figure 2). Nonetheless, these mediators or their precursors must first be secreted to promote any effects.

Since the metabolic abnormalities found in CF homozygote M-Mϕ are expressed in CF heterozygote M-Mϕ as well (Table 5, Figures 13 and 14), the basic genetic defect in CF may be expressed in the M-Mϕ.[24] Further study of the metabolic anomalies underlying CF M-Mϕ hypersecretion and hyperactivity should yield answers which can be applied to exocrine gland dysfunction in general in CF. The molecular events which must occur for secretion to take place are similar in M-Mϕ, PMN, and exocrine cells.[62]

Our studies of the M-Mϕ in CF have already provided several clues to the nature of the metabolic defect(s) underlying CDS accumulation. In addition, our studies with M-Mϕ indicate that CDS accumulation can be corrected *in vitro* and that under certain conditions normal cultures will generate CDS (Table 9). This latter finding is important as it provides the evidence that these mediators may be transiently generated by normal cells and adds strong support to a contention[33] that the CDS are not structurally abnormal products but rather evidence of accumulation of normal components in CF, and that their accumulation does not reflect a defect in gene(s) responsible for dictating their primary protein structure.

SUMMARY

Investigators have been attempting to establish diagnostic criteria for CF and to develop reliable assays for detecting CF heterozygotes for more than 40 years. The most productive effort in this area has been research on biochemical "markers" for the CF gene. Possible applications of knowledge gained from such studies of CF markers include the development of quantitative analytic techniques for detection of CF heterozygotes and also for neonatal and prenatal diagnosis of CF homozygotes. In addition, identification of biochemical markers for the CF gene and understanding of their metabolism could provide insight into the nature of the basic metabolic defect in CF (which is still unknown). Unfortunately, it has proved extremely difficult to find reliable markers for the CF gene. The CDS were first demonstrated in 1967[28] and the CFP in 1973.[11] Only one other potential marker for CF genotypes has been identified, namely, an endopeptidase with trypsin-like activity that is found in normals but not in CF homozygotes.[10,63] This report reviews recent findings on the CFP and CDS and discusses their potential as markers for detecting CF heterozygotes (and ultimately for use in the neonatal and prenatal diagnosis of CF), as well as their importance to our understanding of the basic metabolic defect in CF.

Data compiled from both physiochemical and immunologic studies (Table 8) suggest that CFP may be structurally related to one of the CDS. This is potentially an extremely important observation, since it provides the first evidence that CFP may be synthesized by white blood cells (and possibly fibroblasts), in agreement with our preliminary findings that CFP is present in culture medium from mononuclear leukocytes obtained from CF genotypes. The foregoing observations, combined with the demonstrated specificity of CFP for CF genotypes (Table 3) and data from immunoassays indicating that serum CFP levels are quantitatively different in CF homozygotes (Table 7), seem to indicate that CFP has potential value for prenatal diagnosis of CF homozygotes.

In addition, our data concerning the metabolism of the CDS in M-Mϕ cultures suggest hypotheses about the cellular metabolism of the CDS and their accumulation in the body fluids (particularly serum) of CF genotypes, as well as the possibility that the endopeptidase described by Nadler et al.[10] may play a role in their accumulation. Further studies of CDS metabolism could lead to the identification of even better biochemical markers and hence better assays for the diagnosis of CF and identification of carriers of the CF gene(s).

ACKNOWLEDGMENTS

The expert technical assistance of Eugenia Floyd and Mario T. Parise in executing certain segments of the studies, and the editorial assistance of Charles L. Smith are gratefully acknowledged. The research on the CDS was supported in part by a Basil O'Connor Basic Research Grant, and Administrative Research Grant from the National Foundation, March of Dimes, by grants from the National Cystic Fibrosis Foundation, and by a U.S. Public Health Service Grant.

REFERENCES

1. Fanconi G, Vehlinger E, Knaver C. Celiac syndrome with congenital cystic fibrosis of the pancreas and bronchiectasis. *Wien Med Wochenschr* 1936; 86: 753-63.
2. Andersen DH. Cystic fibrosis of the pancreas and its relation to celiac disease. *Amer J Dis Child* 1938; 56: 344-99.
3. Wood RE, Boat TF, Doershuk CF. State of the art: Cystic fibrosis. *Amer Rev Resp Dis* 1976; 113: 833-78.
4. diSant'Agnese PA, Talamo RC. Pathogenesis and physiopathology of cystic fibrosis of the pancreas: Fibrocytic disease of the pancreas (mucoviscidosis). *N Engl J Med* 1967; 277: 1287-94, 1344-52', 1399-1408.
5. diSant'Agnese PA, Davis PB. Research in cystic fibrosis. *N Engl J Med* 1976; 295: 481-85, 534-41, 597-602.
6. Bowman BH. Factors related to cystic fibrosis. *In:* Mangos JA, Talamo RC, eds. Cystic fibrosis. Projections into the future. New York: Stratton Intercontinental Medical Book Corporation, 1976: 277-84.
7. Lederberg S. Polypeptide mediators of cystic fibrosis. *In:* see Ref. 6. 259-71.
8. Rennert OM. Evaluation of laboratory tests proposed as aids for the diagnosis of cystic fibrosis. *Ann Clin Lab Sci* 1973; 3: 1-12.
9. Bowman BH, Lankford BJ, McNeely C, Carson SD, Barnett DR, Berg K. Cystic fibrosis: Studies with the oyster ciliary assay. *Clin Genet* 1977; 12: 333-43.
10. Nadler HL. Enzyme studies. *In:* see Ref. 6. 285-90.
11. Wilson GB, Jahn TL, Fonseca JR. Demonstration of serum protein differences in cystic fibrosis by isoelectric focusing in thin-layer polyacrylamide gels. *Clin Chim Acta* 1973; 49: 79-91.
12. Wilson GB, Fudenberg HH, Jahn TL. Studies on cystic fibrosis using isoelectric focusing. I. An assay for detection of cystic fibrosis homozygotes and heterozygote carriers from serum. *Pediat Res* 1975; 9: 635-40.
13. Wilson GB, Arnaud P, Fudenberg HH. An improved method for the detection of cystic fibrosis protein in serum using the LKB multiphor electrofocusing apparatus. *Pediat Res* 1977; 11: 986-89.
14. Wilson GB, Fudenberg HH. Studies on cystic fibrosis using isoelectric focusing. IV. Distinction between ciliary dyskinesia activity in cystic fibrosis and asthmatic sera and association of CF protein with the activity in CF serum. *Pediat Res* 1977; 11: 317-24.
15. Wilson GB. Cystic fibrosis protein is a confirmed diagnostic marker for detecting heterozygote carriers. Significance in relation to future screening and to a proposed defect in alpha$_2$-macroglobulin. *Pediat Res* 1979; 13: 1079-81.
16. Pepys MB, Bell AJ, Rowe IR. Sepharose-C3. I. Preparation and use as an immunosorbent. *Scand J Immunol* 1975; 4: Suppl. 3, 75-8.
17. Wilson GB, Fudenberg HH. Further purification and characterization of serum proteins used to detect cystic fibrosis genotypes by isoelectric focusing. *Tex Rep Biol Med* 1976; 34: 51-71.
18. Atland K, Schmidt SR, Kaiser G, Knoche W. Demonstration of a factor in the serum of homozygotes and heterozygotes for cystic fibrosis by a non-biological technique. *Humangenetick* 1975; 28: 207-16.
19. Scholey J, Applegarth DA, Davidson AGF, Wong LTK. Detection of cystic fibrosis protein by electrofocusing. *Pediat Res* 1978; 12: 800-01.

254 Gregory B. Wilson

20. Tulley GW, Nevin GB, Young IR, Nevin NC. Detection of cystic fibrosis protein by isoelectric focusing of serum. *Pediat Res* 1979; 13: 1078.
21. Manson JC, Brock DJH. Development of a quantitative immunoassay for the cystic fibrosis gene. *Lancet* 1980; 1: 330-31.
22. Nevin GB, Nevin NC, Redmond AO, Young IR, Tully GW. Detection of cystic fibrosis homozygotes and heterozygotes by serum isoelectrofocusing. *Hum Genet* 1981; 56: 387-89.
23. Wilson GB. Development of monospecific antisera, hybridoma antibodies, and immunoassays for cystic fibrosis protein. *Lancet* 1980; 2: 323-24.
24. Wilson GB, Fudenberg HH. Does a primary host defense abnormality involving monocytes-macrophages underlie the pathogenesis of lung disease in cystic fibrosis? *Med Hypoth.* In press.
25. Wilson GB, Fudenberg HH, Parise MT, Floyd E. Cystic fibrosis ciliary dyskinesia substances and pulmonary disease: Effects of CDS on neutrophil movement *in vitro. J Clin Invest* 1981; 68: 171-83.
26. Wilson GB, Fudenberg HH. Separation of ciliary dyskinesia substances secreted by cystic fibrosis leukocytes, lymphoid cell lines and found in serum, using protein A Sepharose CL-4B. *J Lab Clin Med* 1978; 92: 463-82.
27. Wilson GB, Jackson HD. Cystic fibrosis: Further characterization of a "new" cytokine which modulates neutrophil and monocyte chemotaxis. *In:* Khan A, Hill NO, eds. Human lymphokines: Biological immune response modifiers. New York: Academic Press. In press.
28. Spock A, Heick HMC, Cress H, Logan WS. Abnormal serum factor in patients with cystic fibrosis of the pancreas. *Pediat Res* 1967; 1: 173-77.
29. Conover JH, Bonforte RJ, Hathaway P, *et al.* Studies on the ciliary dyskinesia factor in cystic fibrosis. I. Bioassay and heterozygote detection in serum. *Pediat Res* 1973; 7: 200-23.
30. Conover JH, Beratis NG, Conod EJ, Ainbender E, Hirschhorn K. Studies on ciliary dyskinesia factor in cystic fibrosis. II. Short term leukocyte cultures and long term lymphoid lines. *Pediat Res* 1973; 7: 224-28.
31. Wilson GB, Bahm VJ. Synthesis and secretion of cystic fibrosis ciliary dyskinesia substances by purified subpopulations of leukocytes *in vitro. J Clin Invest* 1980; 66: 1010-19.
32. Beratis NG, Conover JH, Conod EJ, Bonforte RJ, Hirschhorn K. Studies on ciliary dyskinesia factor in cystic fibrosis. III. Skin fibroblasts and cultured amniotic fluid cells. *Pediat Res* 1973; 7: 958-64.
33. Conover JH, Conod EJ, Hirschhorn K. On the nature of the defect in cystic fibrosis. *Tex Rep Biol Med* 1976; 34: 45-50.
34. Conover JH, Conod EJ, Hirschhorn K. Studies on ciliary dyskinesia factor in cystic fibrosis. IV. Its possible identification as anaphylatoxin (C3a)–IgG complex. *Life Sci* 1974; 14: 253-66.
35. Wilson GB, Fudenberg HH. Studies on cystic fibrosis using isoelectric focusing. II. Demonstration of deficient proteolytic cleavage of $alpha_2$-macroglobulin in cystic fibrosis plasma. *Pediat Res* 1976; 10: 87-96.
36. Hovi T, Mosher D, Vaheri A. Cultured human monocytes synthesize and secrete $alpha_2$–macroglobulin. *J Exp Med* 1977; 145: 1580-89.
37. Mosher DF, Saksela O, Vaheri A. Synthesis and secretion of $alpha_2$-macroglobulin by cultured adherent lung cells. *J Clin Invest* 1977; 60: 1036-45.
38. Wilson GB, Walker JH Jr, Watkins JH Jr, Wolgroch D. Determination of subpopulations of leukocytes involved in the synthesis of $alpha_1$-antitrypsin *in vitro. Proc Soc Exp Biol Med* 1980; 164: 105-14.

39. Wilson GB, Floyd E. Detection and characterization of cystic fibrosis protein employing isoelectric focusing and immunoelectrophoretic techniques. *In*: Allen RC, Arnaud P, eds. Electrophoresis 81. Frankfurt, Germany: Walter de Gruyter 1981; 529-35.

40. Avrameas S, Taudou B, Chuilon S. Glutaraldehyde, cyanuric choride and tetraazotized O-dianisidine as coupling reagents in the passive hemagglutination test. *Immunochemistry* 1969; 6: 67-76.

41. Cambiaso CL, Goffinet A, Vaerman J-P, Heremans JF. Glutaraldehyde-activated aminohexyl-derivative of Sepharose 4B as a new versatile immunoabsorbent. *Immunochemistry* 1975; 12: 273-78.

42. Conover JH, Conod EJ, Hirschhorn K. Ciliary-dyskinesia factor in immunological and pulmonary disease. *Lancet* 1973; 1: 1194.

43. Wilson GB, Monsher MT, Fudenberg HH. Studies on cystic fibrosis using isoelectric focusing. III. Correlation between cystic fibrosis protein and ciliary dyskinesia activity in whole serum shown by a modified rabbit tracheal bioassay. *Pediat Res* 1977; 11: 143-45.

44. Hugli TE. Chemical aspects of serum anaphylatoxins. *Contemp Top Mol Immunol* 1978; 1: 181-214.

45. Ben-Yoseph Y, Defranco CL, Nadler H . The metabolism of sialic acid in cystic fibrosis. *Pediat Res* 1981; 15: 839-42.

46. Wilson GB, Fudenberg HH. Role of polyamines in regulation of proteinase and proteinase inhibitor function: Emphasis on a defect in alpha$_2$– macroglobulin and polyamine metabolism in cystic fibrosis. *In*: Campbell RA, Morris DR, Bartos D, et al., eds. Advances in polyamine research, Vol 2. New York: Raven Press, 1978; 281-305.

47. Hubbard WJ. Alpha$_2$-macroglobulin-enzyme complexes as suppressors of cellular activity. *Cell Immunol* 1978; 39: 388-94.

48. Starkey PM, Barrett AJ. Alpha$_2$-macroglobulin, a physiological regulator of proteinase activity. *In*: Barrett AJ, ed. Proteinases in mammalian cells and tissues. Amsterdam: Elsevier North-Holland, 1977: 663-96.

49. Hubbard WJ, Hess AD, Hsia S, et al. The effects of electrophoretically "slow" and "fast" alpha$_2$-macroblobulin on mixed lymphocyte cultures. *J Immunol* 1981; 126: 292-99.

50. Vischer TL, Berger D. Activation of macrophages to produce neutral proteinases by endocytosis of alpha$_2$-macroglobulin-trypsin complexes. *J Reticuloendothel Soc* 1980; 28: 427-35.

51. Wilson GB. Cystic fibrosis: Evidence for correction of abnormal metabolism of anaphylatoxin (C3a) by monocytes *in vitro* following addition of normal alpha$_2$-macroglobulin complexed with proteases. *Fed Proc* 1980; 39: 451.

52. Morahan PS: Macrophage nomenclature: Where are we going? *J Reticuloendothel Soc* 1980; 27: 223-45.

53. Davies P, Allison AC. Secretion of macrophage enzymes in relation to the pathogensis of chronic inflammation, *In*: Nelson DS, ed. Immunobiology of the macrophage. New York: Academic Press, 1976: 427-61.

54. Whaley K. Biosynthesis of complement components and the regulatory proteins of the alternative complement pathway by human peripheral blood monocytes. *J Exp Med* 1980; 151: 501-16.

55. Ward PA. Chemotactic factors for neutrophils, eosinophils, mononuclear cells, and lymphocytes. *In*: Austen KF, Becker EL, eds. Biochemistry of the acute allergic reactions. New York: Oxford University Press, 1971: 229-42.

56. Laurell CB, Jeppson JO. Protease inhibitors in plasma. *In*: Putnam FW, ed. The plasma proteins. Vol I. New York: Academic Press, 1975: 229-64.
57. Shapiro BL, Feigal RJ, Lam FHL. Intracellular calcium and cystic fibrosis. *In*: Sturgess JM, ed. Perspectives in cystic fibrosis. Toronto: Canadian Cystic Fibrosis Foundation, 1980: 15-28.
58. Martinez JR. Recent advances in cystic fibrosis. *Missouri Med* 1976; 73: 173-78.
59. Galant SP, Norton L, Herbst J, Wood C. Impaired β adrenergic receptor binding and function in cystic fibrosis neutrophils. *J Clin Invest* 1981; 68: 253-58.
60. O'Flaherty JT, Ward PA. Chemotactic factors and the neutrophil. *Sem Hematol* 1979; 16: 163-74.
61. Rao GJS, Chyatte D, Nadler HL. Enhancement of UDPGalactose: Glycoprotein galactosyltransferase in cultured human skin fibroblasts by cationic polypeptides. *Biochim Biophys Acta* 1978; 541: 435-43.
62. Weissman G, Korchak HM, Perez HD, Smolen FE, Goldstein IM, Hoffstein ST. Leukocytes as secretory organs of inflammation. *In*: Weissman G, Samuelsson B, Paoletti R, eds. Advances in inflammation research. New York: Raven Press, Vol. I. 1979: 111-31.
63. Walsh MMJ, Nadler HL. Methylumbelliferylguanidinobenzoate-reactive proteases in human amniotic fluid: Promising marker for the intrauterine detection of cystic fibrosis. *Am J Obstet Gynecol* 1980; 137: 978-982.

GALTON REVISITED

Robin M. Bannerman

The *"englischen Aristokraten"* Francis Galton, as well as the *"oster-reichischen Augustinerpaters"* Johan Gregor Mendel, has long been regarded as a founding father of human genetics.[1] Not all current historians of science agree on his place in history,[2] but this will not be argued here. Nor will any attempt be made to review seriously the two themes in genetics for which Galton is mainly celebrated: biometrics and eugenics. Rather, this discussion concerns Galton as a person, and a Victorian.

The title, *Galton Revisited*, recognizes that all of us interested in this symposium probably have some acquaintance with his work. There are Galton Laboratory alumni present who are doubtless quite familiar with the story, and whom I ask to bear with me. There is no lack of written material on Galton. This short paper owes much to the monumental text *The Life, Letters and Labours of Francis Galton* by Karl Pearson,[3] which was published in three volumes between 1914 and 1930, and to the modern biography by Forest.[4] Galton's own surprisingly brief autobiography is a fascinating book, several times quoted below.[5] Towards the end of his life Galton also wrote a novel. This was never published and, indeed, was suppressed by his family and friends. If this was like many other first novels, it must have contained autobiographical elements and one would wish that it were accessible.

EMINENT VICTORIANS

This heading is taken, of course, from Lytton Strachey's book,[6] one of the sources of insight into that curious era which remains an inescapable part of our present culture. Galton was not one of Strachey's eminent Victorians, but he could well have been, and he could equally have been at home in Edith Sitwell's *English Eccentrics*.

The Victorian era began when Victoria came to the throne in 1837 just after the return of the Beagle. Galton was then 15 years old and later recalled the occasion of the coronation. The relationships between great contemporaries deserve to be more thoroughly pursued. One is tempted to ask, what did each of these know of the others? What did Victoria, whose watchword was "I will be good", think of any of them? She admired Florence Nightingale

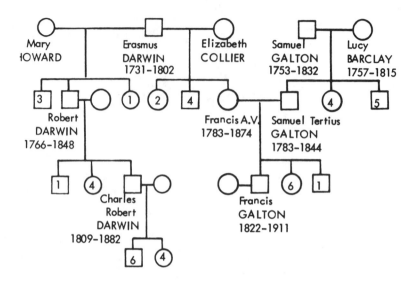

Fig. 1. Pedigree of Francis Galton, from the extensive genealogic data given by Pearson.[3]

TABLE 1
Eminent Victorians

Charles Darwin, 1809-1882
Abraham Lincoln, 1809-1865
Karl Marx, 1818-1883
Victoria Regina, 1819-1901
Florence Nightingale, 1820-1910
Gregor Mendel, 1822-1884
Francis Galton, 1822-1911

and helped her; she was dismayed by Darwin; she was probably aware of Francis Galton but not of his work, and knew of Lincoln as a statesman. But what of Victoria and Karl Marx?

THE GALTON PEDIGREE

Galton came from a large, vigorous and prosperous Midland family (Figure 1). His maternal grandfather was Erasmus Darwin and paternal grandfather Samuel Galton. It is well known that Charles Darwin was his cousin, but less well known that the two were lifelong friends, and had many features in common (Figure 2). Both were physicians *manque*, independently wealthy, given to original thoughts and eccentric behavior, and both have contributed in a pervasive way to knowledge and thinking in the present century.

Little Francis Galton was something of an infant prodigy. He could read at two. At four he wrote his sister "I am four years old and can read any English book. . .and I can read French a little." From such information, Terman later estimated his IQ at 200, for whatever that speculation is worth.[4] While clearly of great intelligence, he was also an attractive child: modest, shy and much liked by people around him. This last quality was remarked upon throughout his life.

GALTON, THE MEDICAL STUDENT

Though the Galton family was rich, they were expected to work. Francis Galton, perhaps surprisingly, was apprenticed to a local physician in Birmingham, and subsequently became an "indoor pupil", at the Birmingham General Hospital. He kept detailed diaries of his experiences and with little training seems to have fulfilled the duties which we would expect of an intern. He greatly enjoyed himself and describes setting fractures, an autopsy in a private home and so on. A medical student's curriculum of the time was more free-floating than ours, and he also studied at King's College London and took time out to travel in Europe before going up to Trinity College, Cambridge in 1840. In a sense, he did part of the clinical before he did the pre-clinical, a procedure which seems to have bothered no one and is worth recalling as we agonize over details of our current medical curricula.

Galton enjoyed Cambridge and made lifelong friends while there, but he was vastly unimpressed by most of the instruction, as he had been in earlier schools. His stay ended apparently in a spell of illness, like that of Charles Darwin, associated with an intermittent pulse and obsessional ideas. He contented himself with a Poll Degree.

Fig. 2a. Francis Galton, aged 38.

Fig. 2b. Charles Darwin at about the same age.

COUNTRY LIFE AND TRAVEL

For a while Galton stayed at home living the life of a country gentleman. In 1844 his father died and he gave up the idea of being a physician since he decided he was sufficiently well off. He wrote "I had many wild oats yet to sow." In the next few years he alternated between country life in England and traveling widely in Africa. He later wrote "It may be thought I was leading an idle life, but not so. I read a good deal all the time and I digested what I read by much thinking about it." How like Charles Darwin's behavior at about the same age when he was the despair of his father to whom he appeared merely idle.

There are several accounts of Darwin's expeditions in Africa, especially his expedition to Damara land. Further quotation from his autobiography[5] is irresistible.

"I did much to make myself agreeable, investing Nangoro with a big theatrical crown that I had bought in Drury Lane for some such purpose. But I have reason to believe that I deeply wounded his pride by the non-acceptance of his niece as, I presume, a temporary wife. I found her installed in my tent in negress finery, raddled with red ochre and butter, and as capable of leaving a mark on anything she touched as a well-inked printer's roller. I was dressed in my one well-preserved suit of white linen, so I had her ejected with scant ceremony. The Damaras are very hospitable in this way, and consider the missionaries to be actuated by pride in not reciprocating."

MARRIAGE AND OTHER INTERESTS

By 1852 Galton found himself somewhat famous through his travels and writings about them, and had by then a wide circle of friends and acquaintances. In 1853, aged 32, he married Louisa Butler, daughter of the Dean of Peterborough. He settled down, to some degree, and his subsequent travels were recreational only. He lived an active social life, among such people as Darwin, Carlyle, Richard Burton and Bishop Wilberforce. He later described the Bishop's efforts to maintain an urbane composure in the face of Richard Burton's uninhibited conversations.

It appears that over these next years he was anxious to do something useful but did not seem to know quite what. For instance, he instructed soldiers in camping methods and he invented the heliostat. He had undertaken, or was in the midst of, a variety of scientific pursuits, including meteorology, geography, photography. Indeed his very versatility may partly explain his controversial place in the history of science. Some of the more curious titles of his works over the years deserve to be cited, and some samples are set out in Table 2.

Fig. 3. Galton the explorer, on his return from Damaraland, aged 33.

Fig. 4. Francis Galton and his wife Louisa, early in their married life.

TABLE 2
Some Galton Titles

On a new principle for the protection of riflemen, 1861
On spectacles for divers, and on the vision of amphibious animals, 1865
Gregariousness in cattle and men, 1871
Statistical inquiries into the efficacy of prayer, 1872
Africa for the Chinese, 1873
Nuts and men, 1874
On the excess of females in the West Indian Islands, 1874
On the average flush of excitement, 1879
Three generations of lunatic cats, 1896

TABLE 3
The Statistical Inefficacy of Prayer (1872)

	Mean age at death
Members of royal houses	64.04
Artists	65.96
Medical men	67.31
Men of literature and science	67.55
Lawyers	68.14
Clergy	69.49
Gentry	70.22

HEREDITY

The publication of Darwin's *Origin of Species* in 1859 marked a turning
point in Galton's life. On reading it, which he apparently did at one sitting, he
"shed the wretched burden of original sin", and from then on worked actively
on biological studies. One of the most celebrated of these was the book *Hereditary Genius*, published in 1869.[7] This was a bad title; Galton meant ability
and he regretted the title all his later life. What he was studying we might now
call the familial aggregation of special abilities. He was attempting to determine
the inheritance of continuously varying traits. This led him on to such problems as height and to the invention of correlation and regression. The book
Hereditary Genius has much of great interest still, and some that is amusing.
Galton was interested in all types of ability, the Bach family, the Bellini
family, English judges, athletes, oarsmen, and cricket players. He also included poets of whom he wrote "high aspirations, but for all that, they are a

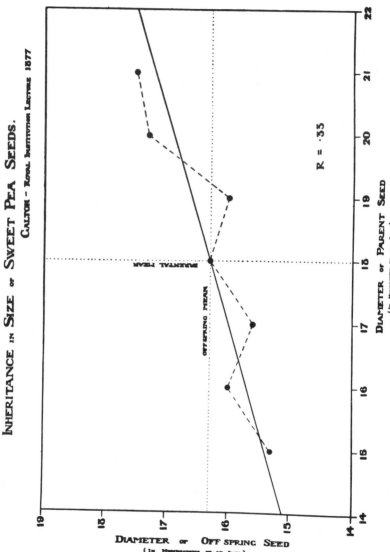

Fig. 5. The first regression line, of size of pea seeds of offspring on parent seed size, 1877.

sensous erotic race, exceedingly irregular in their way of life"; of musicians he wrote, "the irregularity of their lives is commonly extreme". Thus wrote a Victorian! The work received then, and now, a mixed reception. His wife wrote in her diary "Francis' book not well received." Yet Darwin wrote to him, "I do not think I ever in all my life read anything more interesting and original."

A digression that arose from the study of *Hereditary Genius* was the inquiry on the efficacy of prayer, an entirely serious one. The manuscript was rejected by the *Fortnightly Review*, whose editor wrote, "Your paper is too terribly conclusive and offensive not to raise a hornets' nest", but it was later published in the same journal by another editor in 1872. It includes a serious review of the literature on prayer to which was added Galton's own rather original observations. He noted that the most prayed for class in England is the Royal family but found that their average age at death is lower than that of many other classes (Table 3). He also considered the short and miserable lives of missionaries and the fact that insurance companies made no distinction for prayerfulness. He remarked on the commonness of insanity among the nobility, despite the specific prayers for their "grace, wisdom and understanding". This was not irreverently intended. He valued the subjective purpose of prayer, and although at the time an agnostic he was not an unreligious man. He had been a Quaker, and eventually tended towards pantheism. One wonders how this was received by his father-in-law, an Anglican Dean.

Galton's quantitative studies were perhaps his most lasting contribution to the science of human genetics. He published what is probably the first example of a regression curve in 1877 (Figure 5), and, of course, he was the inventor of that word. The data he used in these studies derived from his anthropometric measurements. He invented, for instance, a height predictor (Figure 6) by which the adult height of a child could be predicted from the heights of the parents. In 1884 he had set up an Anthropometric Laboratory. Subjects who came to be examined paid twopence for the privilege, a delightful idea which we might do well to emulate in current studies.

Galton's contributions to human biology are far too numerous even to list in this brief talk, but they included studies of twins and fingerprints. He was greatly interested in the identification of individuals and was one of the people to determine that fingerprints remained the same throughout life. (Figure 7) One of his subjects for prolonged serial observations was the Archbishop of Canterbury. He pressed for the adoption of fingerprints for identification by the police forces in India, Egypt, France — in that order — and finally Scotland Yard itself.

EUGENICS

The idea of eugenics cannot be omitted since it increasingly interested Galton from the 1880s onwards. He coined the word about 1883 and considered three lines of approach. First, he thought that eugenics deserved

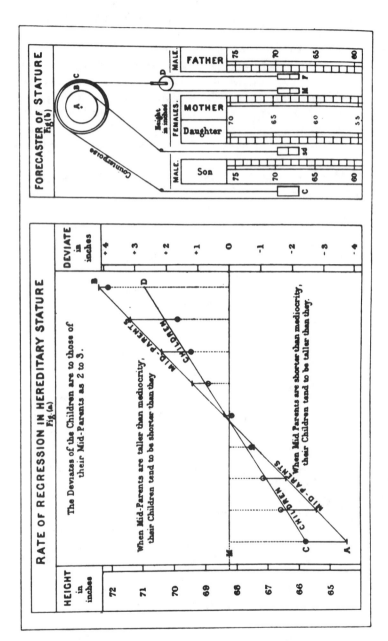

Fig. 6. Galton's height predictor.

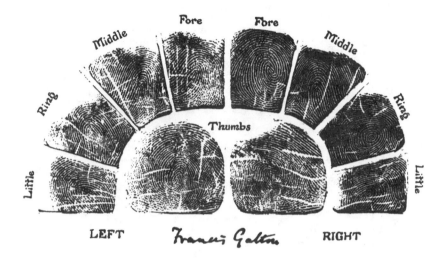

Fig. 7; Francis Galton's fingerprints, 1875, which he observed serially for many years.

Fig. 8. Francis Galton, aged 70.

42, RUTLAND GATE. *Dec.* 19 75.

MY DEAR DARWIN,   The explanation of what you propose does not seem to me in any way different on my theory, to what it would be in any theory of organic units.   It would be this:

Let us deal with a single quality, for clearness of explanation, and suppose that in some particular plant or animal and in some particular structure, the hybrid between white and black forms was exactly intermediate, viz: grey—thenceforward for ever. Then a bit of the tinted structure under the microscope would have a form which might be drawn as in a diagram, as follows:—

whereas in the hybrid it would be either that some cells were white and others black, and nearly the same proportion of each, as in (1) giving *on the whole* when less highly magnified a

(1)   (2)

uniform grey tint,—or else as in (2) in which *each cell* had a uniform grey tint.

Fig. 9. How to explain a blended appearance in a hybrid; Galton to Darwin, 1875.

study and for this purpose he was instrumental in founding the journal *Biometrica* in 1901. He created a Research Scholarship in 1905, and the Eugenics Laboratory, later the Galton Laboratory, in 1909. Second, he thought that eugenics should be recognized as being of social importance and for this purpose he helped to found the Eugenics Society. Third, he felt that eugenics should be introduced as a "practical religion", and it was this approach which led to the greatest subsequent trouble. Galton's original conception of eugenics was positive. He intended to encourage procreation by the most healthy and appropriate parents. He was well aware of the presumption implied in the choice of parents and quite humble about this. His aim was not to seek a superior group but to represent each class or sect, to achieve variety not uniformity, and to consider physique, ability and character. The Eugenics Society was open to everyone and soon led to considerable difficulties so that he was led in 1909 to publish his essays in eugenics in which he reiterated that positive eugenics was more important than negative. The subsequent destructive divergence of the eugenic movement is a history in itself and only now is it possible to discuss the topic again.

THE END OF THE ERA

Queen Victoria died in 1901. Galton outlived her and in 1908 published his autobiography, *Memories of My Life,* which was as much concerned with travel, exploration and hunting as it was with heredity and eugenics. The controversy between the Mendelists and the Galtonists was already in progress. In the autobiography he discussed his relationship with Mendel, for whom he had a great fondness, partly from the chance of their having been born in the same year. One has the impression that the conflict between the Mendelists and the Galtonists was none of Galton's doing and that he was in his heart a Mendelist. Years before, he and Charles Darwin had corresponded about the explanation of hybrids. Did they represent a blend of two extremes, or, at the cellular level was there a mosaic? A section of his correspondence is shown here in Figure 9.

To the end of his life, Galton was well loved, revered, and perhaps not always understood. I shall leave the last word with the redoubtable Florence Nightingale, who wrote, apparently of Galton, as cited by Pearson:[8]

"But all these must be splendid Madmen who initiate any great theory, any great work which does not recommend itself to the present knowledge or ignorance of minds which do not see so far as the splendid madmen of this Age, who will be sensible men to the next Age, and perhaps a little in arrear to the Age after that."

ACKNOWLEDGMENT

Figures 2 through 9 have been reproduced by courtesy of the Cambridge University Press (see Ref. 3).

REFERENCES

1. Vogel F. Lehrbuch der Allgemeinen Humangenetik. Berlin: Springer, 1961.
2. Cowan RS. Francis Galton's contribution to genetics. *J History Biol* 1972; 5: 389-412.
3. Pearson K. The life, letters and labours of Francis Galton. 3 Vols. London: Cambridge University Press, 1914-30.
4. Forest DW. Francis Galton. The life and work of a Victorian genius. New York: Taplinger, 1974.
5. Galton F. Memories of my life. London: Methuen, 1908.
6. Strachey L. Eminent Victorians. London: Chatto and Windus, 1918.
7. Galton F. Hereditary genius. 2nd ed. 1892. Cleveland: World Publishing Company, 1962.
8. Pearson K. Francis Galton. 1822-1922. A centenary appreciation. Eugenics Laboratory Publications. Questions of the day and fray No. XI. London: Cambridge University Press, 1922.

INDEX